MISSISSIPPI RIVER TRAGEDIES

MISSISSIPPI RIVER TRAGEDIES

A Century of Unnatural Disaster

CHRISTINE A. KLEIN

SANDRA B. ZELLMER

NEW YORK UNIVERSITY PRESS
New York and London

NEW YORK UNIVERSITY PRESS
New York and London
www.nyupress.org

References to Internet websites (URLs) were accurate at the time of writing. Neither the author nor New York University Press is responsible for URLs that may have expired or changed since the manuscript was prepared.

For Library of Congress Cataloging-in-Publication data, please contact the Library of Congress.
ISBN 978-1-4798-2538-7 (hardcover)

New York University Press books are printed on acid-free paper, and their binding materials are chosen for strength and durability. We strive to use environmentally responsible suppliers and materials to the greatest extent possible in publishing our books.

Manufactured in the United States of America

10 9 8 7 6 5 4 3 2 1

Also available as an ebook

To Randy, ever patient and supportive.
SBZ

To Mark, with love always.
CK

CONTENTS

The Mississippi Basin

Source: National Park Service

———•———

MISSISSIPPI RIVER CHILDREN

The Headwaters: Notes from Sandra B. Zellmer

When I was little, my mother bathed me in a garbage can filled with Mississippi River water. Not every night, of course, but just about every summer when my family was camping near the river's headwaters in northern Minnesota. I suppose I smelled a little fishy, but the aroma of river water was completely familiar—and comforting—to me. I savored the names of the headwater lakes where we camped, titles bestowed by Chippewa and Dakota Indians or by European explorers: Itasca, Winnibigoshish, Andrusia, Bemidji, LaSalle.

My passion for the outdoors comes naturally. My father was a third-generation German American farmer who raised cattle, corn, and alfalfa just outside of Sioux City, Iowa, nestled in the valley of the Missouri River, the longest tributary of the Mississippi. He was following in the footsteps of my great-grandfather Gustav Zellmer, who arrived at the Castle Garden Immigration Depot in New York in 1883, straight off the boat from Kolmar, Germany (now part of Poland). Sixteen-year-old Gustav was anxious to make his mark on the New World. He rode the trains west, marveling as he crossed over the Mississippi River and entered Iowa. He stopped when he reached the Missouri River. He had never seen such black, fertile soil, and he was awed by the gently rolling terrain, perfect for the plow. The rich dirt came at a price, however. It was formed and nourished by floods, like the one in 1892, which made Gustav's house list to one side and float away from its foundation. The family escaped—their baby daughter, my great-aunt Henriette, was carried to safety by the town doctor—and the house itself was later moved to higher ground.

By the time of the Great Depression of the 1930s, Gustav had amassed quite a bit of prime farmland, as well as eight children to help him. His youngest son, my grandfather (also named Gustav), relished everything the elements could throw at him. He was willing to take calculated risks and to ride out the bad years while waiting for the good ones. Most of the time, Grandpa Gus' bets on corn and cattle paid off, despite floods, tornadoes, drought, hail, and pests.

My father, Mervin, saw things a little differently. As the smartest boy in his high school class, he dreamed of going to college. When he met my mother at a dance in the river bottoms of Hornick, Iowa, he was drawn to her brilliant smile and even more to her quick wit and her own bold dreams for the future. Then the Army called with other plans.

Mervin and Jessie Zellmer were married on May 27, 1951, just a few weeks before Mervin reported to boot camp. They took a weeklong honeymoon to a magical place that my father had discovered a few years earlier on a fishing trip with a buddy—Lake Itasca and the other headwater lakes of the Mississippi River. Instead of being squeamish like most girls he knew, my mom took to fishing and to the north woods as if she were born to it. They chased each other over the stepping stones that crossed the headwaters of the Mississippi, rented a small boat, and snapped photographs of their adventures fishing for walleye. Mom caught the prize-winner—a hefty twelve-pounder.

By June 1951 when my father arrived at Fort Riley on the Kansas River (a tributary of the Missouri), it had been raining steadily for nearly two months. In reel-to-reel tapes he recorded for my mother, Dad reported that late one night in the barracks he awoke to water lapping up beside his cot. All of the men of his unit were ordered to pack up and seek higher ground. The barracks were destroyed. For the rest of my father's eight-week tenure at Fort Riley, the men slept in pup tents on a ridge. Years later, Dad reminisced about the incessant rain, and told us stories about the poisonous snakes and saucer-sized spiders that sought higher ground, too. The soldier who forgot to shake out his boots in the morning was often very sorry for his carelessness.

As it turned out, that summer brought one of the worst floods in the region's history. During a four-day period in early July, up to sixteen inches of rain fell on already-saturated soils. On just one day—July 13, 1951—floodwaters crept across nearly two million acres in Kansas and Missouri.

Dad eventually came home from the Korean War, unharmed, and in 1956 my parents bought a farm of their own near Sioux City. Farming in the 1950s was challenging at first, and not terribly profitable. There was not much time to travel, so my family stuck close to home and explored the Missouri River on weekends. My older sisters recall making a family decision while sitting around the dinner table one evening before I was born—whether to purchase their first color television set or a motorboat. It was unanimous. They chose the boat, a sixteen-footer called "Old Blue."

By the mid-1960s, the family's weekdays had settled into the rhythms of farm life: planting, cultivating,

Jessie Zellmer with her monster walleye,
Itasca State Park, 1951
Photograph by Mervin Zellmer

harvesting, and tending livestock. On weekends, we went boating (Old Blue lasted well into the 1980s), fishing, mushroom-hunting (morels being only somewhat less elusive than walleyes), and camping along the Missouri River. Whether we went upstream into South Dakota or downstream toward Omaha, the river was different every time we ventured out. New sandbars and beaches appeared in unexpected places, while others that we had picnicked on just the previous weekend vanished without a trace. I learned to appreciate the river's mercurial nature and its power when I watched it sweep away my sister Marnie and my cousin Susan, who were playing in the current just a few yards from shore. As much as they hated wearing lifejackets, this particular story might not have had a happy ending if they had gone without.

Despite the delights of the Missouri, my parents never forgot about the headwaters of the Mississippi River. Starting with the summer before my second birthday, when farming had begun to pay off, we frequently made the eight-hour drive up to Lake Itasca and nearby lakes and tributaries. According to family lore, I learned to swim before I could walk. It was not long before I was catching crappie, bass, northern pike, and, with a lot of luck, walleye. If I caught fish, I was expected to clean them—all except walleye, whose flesh was far too precious to sacrifice to clumsy young fingers. The year I graduated from high school, my mother invested her nest egg in a small cabin on Long Lake (one of the ten thousand lakes boasted of by Minnesotans, dozens of which are named "Long"), and we settled down as part-time lake residents.

My mom passed away just a few years before we began writing this book, and my father followed soon after. I am only twenty miles from their honeymoon spot as I write this passage at the cabin I inherited. Listening to a pair of loons and looking out at the rain on the lake, I feel them here, sitting beside me at the kitchen table. Their lessons have stuck—the power of water, the beauty of the creatures that occupy the rivers, the fertility of the floodplain, and the measure of independence and self-reliance that could only come (for me, at least) from a childhood spent outdoors. I guess you could say there's something in the water, because three of my nephews are farming, and like me they would rather be outside than anywhere else.

Downstream in St. Louis: Notes from Christine A. Klein

"WAIT, MAGGIE," I called out to my friend, as my preteen legs pumped hard on the pedals of my new Schwinn. The bicycle was a beauty—a full sixteen-inch adult frame, in striking lime green. I had just bought it with $120 of hard-earned babysitting money. I could barely reach the pedals, but I would grow into it, my parents assured me, with the kind of Midwestern thriftiness that had no use for a succession of quickly outgrown bicycles. On that muggy August morning in the late 1960s, Maggie and I were whizzing down the steep curves of Hog Hollow Drive in Chesterfield, Missouri, a suburb of St. Louis. We were savoring the last days of summer as we sped toward the broad valley that bordered the Missouri River, just above its confluence with the Mississippi.

The river bottom was one of my favorite places. I spent a lot of time there, deterred only when the river swelled over its banks and officials stretched a "road closed due to flooding" sign across the top of Hog Hollow Drive. When dry, the area was ideal for bicycling —flat and straight, except for the brief, steep stretch of Hog Hollow that connected

Looking down Hog Hollow Drive
Photograph by Christine A. Klein, 2012

the river valley to the neighboring uplands. It was also prime territory to explore with my dog, Ginger. There was very little traffic to worry about. It fact, there was very little of anything other than wide open farmland and the county waterworks facility—a collection of gated holding ponds and impenetrable-looking buildings that emitted a low-pitched drone. The humming emptiness was eerie, adding welcome mystery to our lives.

We frequently ducked under the plant's gated barrier, ignoring the "no trespassing" signs, and crunched down the waterworks gravel road to the banks of the Missouri River. There, we were rewarded with the sight of dark waters flowing swiftly to places undoubtedly more glamorous than those we knew. I was careful to restrict my dog's swimming to calm eddies, safe from the powerful current of the main channel. Ginger and I invariably returned home caked with the shoe-sucking, silty muck known as *gumbo mud*. Its smell reminded me of warm sunlight and frogs. Back home, my mother banished us to the garage where, armed with a bucket of warm water and towels, I scrubbed myself and Ginger until we could pass my mother's rigorous inspection for lingering traces of the river.

Maggie and I were great admirers of two other Mississippi valley natives, Tom Sawyer and Huck Finn. We decided to build a raft, with vague plans to float down the Missouri and Mississippi rivers in the wake of Tom, Huck, and numerous other intrepid explorers we had studied in school. Noting the buoyancy (and easy availability) of tin cans, we nailed several dozen of them to a platform pieced together from wood scraps, naively confident that the awkward mass would float. We tested our creation one January day on a small lake that had not yet iced over. Ginger and I climbed aboard the craft, and Maggie pushed us out. We promptly sank. Fortunately, the lake was quite shallow and I was able to slosh back to shore with Ginger. We walked home, shivering, where my mom set us up for our usual garage cleanup routine.

I wasn't the only one in my family who was drawn to the river bottoms. On many summer mornings after Sunday church services, my

The turnoff to the Waterworks, with Hog Hollow Drive rising from the
floodplain in the background

Photograph by Christine A. Klein, 2012

dad drove us down Hog Hollow Drive to Rombach Farms' produce
stand. There, we bought just-picked tomatoes, sweet corn, and cantaloupe. We particularly savored the sun-warmed tomatoes, with a
mouth-filling taste that I still associate with black river soils. On the
drive home, the mixed scents were intoxicating, as barely acidic whiffs
of tomato cut through the cantaloupe's musky sweetness.

Just as I was a product of rivers, so, too, was my hometown. St.
Louis was established in 1763 as a trading post, strategically located
where the Missouri and Mississippi rivers came together. In 1817, the
city welcomed its first steamboat, the *Zebulon M. Pike* (named after
the nineteenth-century explorer), and by later in the century, St. Louis
was one of the nation's largest ports, second only to New York and
New Orleans. This commercial success attracted mid-century waves of

German and Irish immigrants. My great-grandfather Herman Flebbe was one of them, leaving Larstedt, Germany, in 1871. As a young man, he established the Western Candy and Bakers' Supply Company, distributing flour, sugar, and other raw materials to confectioners in St. Louis and points beyond. I still have a marble-topped ice cream parlor table from his showroom.

To celebrate the city's history as the "gateway to the West," the 630-foot tall St. Louis Gateway Arch was built downtown in the Mississippi Riverfront district in the mid-1960s. The engineers called for the simultaneous construction of both legs of the Arch, confident that they would meet precisely as calculated to form one continuous curve. The margin of error was only 1/64 of an inch. My family made frequent trips to the riverfront where we would lie on the grass, squint up, and worry whether the legs would actually intersect high above. Remarkably, they did. We were among the first passengers to ride up the tram to the top of the Arch to take in the view, thrilled as the structure swayed several inches in the wind. The Mississippi River was just to the east. Eleven blocks due west, we could see the spot where Great-Grandpa Flebbe's business had been. Farther west, barely out of sight, lay the Missouri River bottoms and Hog Hollow Drive.

Through Adult Eyes

OUR CHILDHOOD PASSIONS for rivers inspired this book. We both grew up in the Mississippi River basin, during a period when kids had plenty of freedom to play and explore outside. Our early lives were infused with the sights, smells, and feel of rivers and their soils. Although we both became law professors, with specialties in water and other natural resources, our interest in the Mississippi River remains personal. As we chronicle the basin's history over the past century—revealing a tug-of-war between the river's natural inclinations and society's desires and laws—we draw on our early memories for inspiration. As we look back on our childhoods, we are faced with many questions.

Why were the places so special, the soils so rich, the flooding so pervasive? What were the biggest disasters in the Mississippi River basin over the past century, both natural and otherwise, and what caused them?

When Mervin and Jessie Zellmer honeymooned at Lake Itasca in 1951, they were enjoying one of the least altered lakes of the Mississippi headwaters, in large part because Minnesota had designated the area a state park. But all around them, dams had been constructed on six other lakes just below Itasca to support navigation and to control flooding. As much as they loved the natural beauty of the north woods, the newlyweds, like most Americans of their generation, applauded the feats of engineering designed to protect them from floodwaters. Was their admiration warranted?

When Great-Grandpa Flebbe established his business near the banks of the Mississippi, he benefited from navigational improvements that helped make St. Louis a vibrant commercial center. Over the years, such engineering efforts, together with economic development, crept west to the Missouri River. Today, swaths of the fertile floodplain lie beneath layers of asphalt. Although Rombach Farms still remains, developers paved over many other farms, making way for business centers that include the country's largest strip mall. Do the economic revenues justify the loss of rich farmland and river habitat?

As we sift through the loam of our childhoods, recalling our early intuitions about rivers, we ponder these questions. In the end, it all comes down to the blurry line between natural and unnatural disasters, and the law's ability to anticipate and respond to them.

ACKNOWLEDGMENTS

———•———

CHRISTINE A. KLEIN thanks the University of Florida Levin College of Law for a generous sabbatical research grant. Peter Morris contributed invaluable research assistance and Tim Meyer was a master fact-checker (any remaining errors, of course, are the responsibility of the authors). The writers' group in Boulder, Colorado, was a wonderful sounding board for early drafts of this book, and the author is grateful for the patience and wisdom of Stephanie Bendel, Joanne Brothers, John Christenson, Judy Gilligan, Susan Solomon, and Joe White. My dear friends Cynthia Barnett and Liz Knapp tirelessly slogged through an early draft of this manuscript, with no comma too insignificant to capture their attention. As always, my husband, Mark Ely, was my best supporter, and never became impatient with my endless requests to review each chapter "just one more time."

* * *

SANDRA B. ZELLMER thanks the University of Nebraska College of Law for a generous summer research grant, and Samantha Pelster, Emily Rose, and Patrick Andrews for their top-notch investigative contributions. Randy Mercural, my husband, has my deepest gratitude for his love, encouragement, and good humor throughout this book project.

INTRODUCTION

DISASTERS, NATURAL AND OTHERWISE

Drive through any suburban area and you are likely to find subdivisions with names like "Oak Tree Farms," "Meadow View," and "Eagle's Nest." But try to find the features that inspired those names, and you may discover that the trees, meadows, and nests have given way to farms, neighborhoods, and lush lawns. Are those places still "natural," even though sod has replaced meadow, and dog houses have replaced bird nests? Walk into any grocery store and there will probably be an aisle dedicated to natural foods. Does that suggest, somehow, that the stock filling the rest of the aisles is "unnatural"?

The fuzzy line between natural and unnatural reflects ambivalent attitudes toward nature. We idealize it, naming our neighborhoods and our healthiest foods in its honor. And yet we also see nature as an adversary to be conquered, blaming it for such "natural disasters" as floods, storms, hurricanes, and erosion. Sometimes, we even blame the Almighty and attribute our woes to "acts of God."

Nowhere is this tension clearer than in the Mississippi River basin. The great river and its tributaries flow through, drain, or form the border of more than thirty states. Overall, the Mississippi drains about 40 percent of the continental United States, from Montana to New York, from New Mexico to North Carolina, and from Minnesota down to

Louisiana. The U.S. Army Corps of Engineers, the federal agency in charge of managing the river, describes it as one of the nation's "outstanding assets." But the Corps also asserts that the Mississippi, in its natural condition, represents a "liability . . . [that poses] a threat to the security of the valley through which it flows."[1]

When calamity strikes in the Mississippi basin, our first impulse is to shudder at the uncontrollable fury of nature. We sense, deep in our gut, that it was only a matter of time before the Mississippi unleashed a natural disaster, revealing itself as the deadly liability recognized by the Corps. And what's worse, we fear that we have no control over the disaster and that we are powerless to stop it. Nothing could be further from the truth.

War in the Mississippi Basin

THE MISSISSIPPI RIVER flows through one of the most highly engineered river basins in the world. Today, if you were to fly over the river, you might think that the upper Mississippi was not a river at all, but rather a chain of large lakes, one thousand miles long and as much as three miles wide. Concrete chambers—*locks*—punctuate the upper Mississippi, serving as a watery staircase that allows boats of all shapes and sizes to navigate the river's uneven course. Crafts headed downstream wait in one lock as the dam opens and water drains into the lock below, and then continue on their journey when the levels are equalized. To travel upstream, the process is reversed: boats wait in the lower lock, floating up as dam-released water flows in from above. There are twenty-nine pairs of such locks and dams on the upper Mississippi, extending from northern Minnesota past St. Louis to the mouth of the Ohio River at Cairo (pronounced *Kay*-roh), Illinois. This river segment has been transformed so dramatically that it resembles a set of steps more than a natural water body. The architect of the transformation, the Army Corps of Engineers, refers to its handiwork as a "stairway of water."[2]

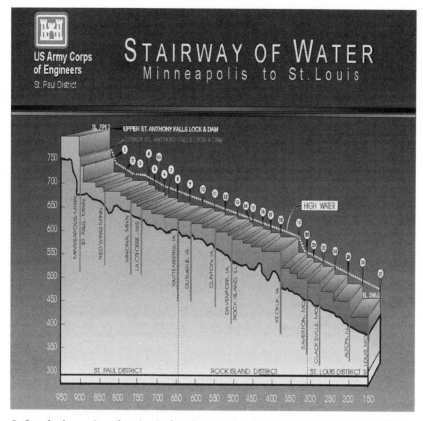

Left scale shows river elevation in feet above sea level; bottom scale shows river miles to
the confluence with the Ohio River at Cairo, Illinois

Source: U.S. Army Corps of Engineers

The locks are a marvel of modern engineering. But even before they
were built, engineers had attempted to tame the river by dredging mud
and silt from its channels and by blanketing its shoreline with levees,
floodwalls, jetties, and other structures designed to control floods. Now,
a 1,607-mile levee system lines the lower Mississippi River, from Cairo
all the way downstream to the Gulf of Mexico. An additional 596 miles
of levees extend along southern tributaries of the river.[3]

A bird's-eye view of the Mississippi delta, where the river meets the
gulf, reveals multiple hues of blues and browns, where freshwater mixes

with seawater and where silt, sand, and clay are deposited, layer by layer, creating side streams called *distributaries* that carry water and sediment to the ocean. In addition to these natural distributaries are channels that have been dredged into the delta to promote shipping and oil and gas development. Situated between the distributaries and channels are low-lying pockets of land created from river deposits—bayous, marshes, and coastal wetlands—that look like the webbing of a duck's foot.

The Corps of Engineers has struggled mightily to control the Mississippi—an effort it likens to war. The metaphor is not surprising, given the Corps' military pedigree. On the eve of the American Revolution, the Continental Congress established not only the Army, but also named a chief of engineers. (Colonel Richard Gridley, appointed in 1775, was the first.) Since that time, the chief and his Corps—made up of both military and civilian personnel—have provided engineering support for military and civilian matters. In the Corps' words, its mission is to "[p]rovide vital public engineering services in peace and war to strengthen our Nation's security, energize the economy, and reduce risks from disasters."[4]

The Corps takes the risk-reduction aspect of its mission seriously, particularly when it comes to the Mississippi basin. In vivid prose, unexpected in a bureaucratic document, the Corps' Mississippi Valley Division describes the focus of its work as the "contumacious" Mississippi River. Explaining the difficulty of its task, the Corps refers to the river as both "beast" and "benefactor": "This Janus-faced colossus periodically seeks to challenge the flood control system imposed upon it, while its opposite profile is a vital waterway network that extends into the heart of the nation—a true cornerstone of our economy."[5]

"Contumacious"? "Janus-faced colossus"? The Army does not mince words—or tread lightly—when it comes to battle with what it perceives as the stubborn and willfully disobedient river. Clearly, the Mississippi River has been modified by many human hands, including those of the Corps, following the instructions of Congress, responding to the will of the electorate. But just as clearly, some of those efforts have backfired.

Law and Unnatural Disasters

TO EXAMINE THE relationship between human action and disaster, legal scholars have called for the development of a new area of study. In 2006, professors Daniel A. Farber and Jim Chen published *Disasters and the Law: Katrina and Beyond*, a law school textbook that considers legal rules that deal with catastrophic risks, including prevention, insurance, emergency response, compensation, and rebuilding strategies. As the authors explained, "we are all stunned by each new disaster, but rapidly come to view it as exceptional and never to be repeated. Thus, we fail to prepare for the next one."[6] Instead of this insufficient, piecemeal response, the authors highlight the need for a comprehensive legal approach to major disasters. The developing field has come to be known as *disaster law*.

The systematic study of disaster poses intellectual puzzles that involve law and a variety of other academic disciplines. One of the thorniest questions—and one of the main themes of this book—is where to draw the line between "natural" and "unnatural" disasters. This challenge was taken up by historian Ted Steinberg in *Acts of God: The Unnatural History of Natural Disaster in America*. Steinberg traces the practice of blaming nature for calamity to the late nineteenth century, when the wide-spread belief that disasters were God's punishment for sin gradually gave way to the notion of nature as culprit in a shift that neatly excused humans from moral or other accountability for harm. As Steinberg argues, "This constrained vision of responsibility, this belief that such disasters stem solely from random natural forces, is tantamount to saying that they lie entirely outside human history, beyond our influence, beyond moral reason, beyond control."

Blaming nature also proved to be politically expedient, as powerful figures sought to "normalize calamity." By convincing us that we should expect random strikes of nature, Steinberg concludes, our leaders "have been able to rationalize the economic choices that help to explain why the poor and people of color—who have largely borne the brunt of these disasters—tend to wind up in harm's way."[7]

The Law Falls Short: A Brief Detour
outside the Mississippi Basin

TODAY, HYDROLOGISTS AND hydrogeologists can tell us much about the movement of water. Their study includes runoff processes, stream-flow routing, and flood frequency analysis. The scientists examine the relationship between rivers and groundwater. Moreover, they develop complex computer models to describe the action of water under a variety of scenarios. The terminology can be daunting to a layperson: *unconfined aquifer, hydraulic conductivity, transmissivity, Darcy's law, evapotranspiration.*

Overall, scientists have learned that a variety of factors—both natural and human—can influence the behavior of water. Before all this study, early judges were perplexed by the movement of water. When it came to underground water, in particular, many threw up their hands in befuddlement. As one Connecticut judge explained in 1850, "The laws of [groundwater's] existence and progress . . . cannot be known or regulated. It rises to great heights, and moves collaterally, by influences beyond our apprehension. These influences are so secret, changeable and uncontrollable, we cannot subject them to the regulations of law, nor build upon them a system or rules, as has been done with streams upon the surface."[8]

Disputes over surface water were a little easier for judges, but even so, they struggled to sort out the natural and human causes of flooding. Under one popular legal theory—the *common enemy* doctrine—judges recognized a common need to vanquish floodwaters, and refused to hold anyone accountable for redirecting surface flow and incidentally drowning out the neighbors. This theory took to heart the ancient Latin maxim that translates as "To whomsoever the soil belongs, he owns also to the sky and to the depths." Under this view, landowners could manipulate the water and soil of their property with impunity, from the bowels of the earth to the heavens above. For example, in one case, a property owner changed the drainage pattern of his land to such an extent that it caused water to back up inside his neighbor's house, to overcome its septic system, and to fill the house and yard with three feet of water and the pervasive odor of raw sewage. Still, recognizing flood water as an enemy to all, the court refused to hold the first landowner responsible, even though the sequence of cause and effect was clear.[9] A second approach, known as the *natural flow* rule (or the *civil law* rule), went to the opposite extreme. Instead of encouraging a mad scramble of uncoordinated drainage measures, it cut off self-protection with the threat of legal liability. Under this rule, landowners could not obstruct the natural flow of surface water, including floods, by any means whatsoever if such efforts would harm their neighbors.[10]

Over time, judges recognized the impracticality and unfairness of both theories. Increasingly, they cast aside both the common enemy and natural flow rules, noting their "anarchic" nature and their "deplorable rigidity," respectively. In the words of one judge, in a mature economy it makes little sense that land development costs "should be borne in every case by adjoining landowners rather than by those who engage in such projects for profit."[11] As a replacement for the older rules, judges turned to the compromise *reasonable use* standard, which holds landowners accountable for harm to others, but only if their interference with the flow of surface waters and use of their property is "unreasonable." Under this approach, a mere cause-and-effect relationship between action and harm is not enough to trigger legal liability. Overall, jurists praise the rule of reasonableness for its fairness and flexibility. But the distinction between "reasonable" and "unreasonable" actions is notoriously squishy, and plagues numerous legal disputes well beyond the context of flooding. Where should courts draw the line? As one judge explained as early as 1894, "In a philosophical sense, the consequences of an act go forward to eternity, and the causes of an event go back to the dawn of human events, and beyond. But any attempt to impose responsibility upon such a basis would result in infinite liability for all wrongful acts."[12]

The core problem is this: we want to extend responsibility far enough to deter or punish socially undesirable actions, but not so far that the result is unfair or unworkable. It's a tough challenge. Just ask any law student who has studied the 1924 case of the unfortunate Mrs. Palsgraf.

* * *

MRS. HELEN PALSGRAF was injured in 1924 at a New York railroad station through an unusual turn of events: A tardy passenger attempted to board a moving train. Two helpful railroad employees reached out to assist, accidentally causing the passenger to drop a newspaper-wrapped parcel. The package turned out to contain fireworks, which exploded on impact with the ground. The resultant vibrations knocked over several large scales used for weighing luggage, which were located at the other end of the railroad platform. They tumbled onto the unsuspecting Mrs. Palsgraf as she waited for her train more than twenty-five feet away from the late-arriving passenger. Basing her suit on the somewhat unlikely chain of events that had unfolded, Mrs. Palsgraf sued the Long Island Railroad Company for her injuries.

There was no question that Mrs. Palsgraf had in fact been hurt that day. It was also clear that she would not have been injured in the absence of the railroad employees' actions. Despite these factors, the New York court ruled against her in *Palsgraf v. Long Island Railroad Company*. Eminent jurist Benjamin Cardozo, joined by three other judges, emphasized that not all actors who cause harm bear legal responsibility. Articulating the often-repeated *zone of danger* rationale, Cardozo asserted that although the railroad employees reasonably could have anticipated that shoving the passenger might hurt *someone* (most likely, the passenger himself), they could not have anticipated that dislodging the passenger's innocuous-looking parcel posed a risk of harm to Mrs. Palsgraf in particular, who was standing many feet away.

Judge Andrews and two other judges disagreed with Judge Cardozo. In their dissenting opinion, they argued that the railroad should have been held accountable as the *proximate cause* (or legal cause) of the accident because there was "a natural and continuous sequence between cause and effect" and thus, it was "reasonably foreseeable" that someone would have been injured by the shove, even if the precise sequence of events could not have been predicted. Today, judges regard both tests—zone of danger and foreseeability—as useful tools for sorting out who should be held accountable for accidents. Almost a century after Mrs. Palsgraf's case, these doctrines are still used to determine legal responsibility for harm of all sorts, including flooding in the Mississippi basin, as described in subsequent chapters.

Fire in the Gulf of Mexico

OLD HABITS DIE hard. In particular, it's tough to stop thinking of all watery catastrophes as "natural" events, or at least as unpredictable accidents. In April 2010, the *Deepwater Horizon* oil-drilling rig exploded in a ball of flame. Floating in almost a mile of water, the rig was just beyond the continental shelf and about forty miles off the coast of Louisiana where the Mississippi River dumps into the Gulf of Mexico. Slowly, the rig sank to the bottom. Oil gushed from the Macondo well and fouled the gulf and its wildlife while British Petroleum, the well's owner, tried one unsuccessful fix after another. BP fumbled for almost three months before it was able to stem the flow of oil. All the while, the world watched real-time undersea videos as BP tried to maneuver well-sealing equipment precisely into position.

Clearly, the "experts" were unprepared for such a disaster. As part of its application for a federal drilling permit, BP submitted an oil spill response plan. Among other things, the plan discussed potential impacts to walruses. Embarrassingly enough, walruses had not graced the waters of the gulf for some three million years. The response plan, it appears, was a cut-and-paste effort that incorporated risk analyses from applications to drill in Arctic waters, where walrus populations can indeed be found. The plan also listed the telephone number of Peter Lutz, a wildlife specialist in Florida who could provide advice in the event of an emergency. As it turned out, Mr. Lutz would be unable to offer assistance during the 2010 blowout because, unfortunately, he had passed away in 2005. But BP alone cannot be blamed for the bumbling disaster plan. Later, a congressional panel determined that Exxon Mobil, Chevron, and Conoco Phillips all relied on similarly flawed contingency plans for drilling in the gulf. With each one citing the need to protect walruses, and several of them providing the phone number of the late Mr. Lutz, these plans were derided as "cookie cutter" by the panel.

Nothing about the circumstances of the BP fire was natural. The rig was a behemoth of four decks stacked one above the other, topped off by a twenty-five-story oil derrick. The national commission that investigated the disaster, moved beyond dry, bureaucratic reporting, described how "the derrick fire roared upward into the night sky, an inferno throwing off searing heat and clouds of black smoke. The blinding yellow of the flames was the only illumination except for the occasional flashlight." In all, eleven people died. Others escaped in the rig's lifeboats, lowering themselves down 125 feet from the rig to the gulf.

Afterward, there was plenty of blame (and legal liability) to go around. The national commission found fault with BP (as owner of the Macondo well), with the Halliburton Company (as the contractor that provided the well casing), and with Transocean, Ltd. (as owner of the drilling rig). In addition, the commission faulted the federal oversight agency formerly known as the Minerals Management Service for inadequate regulation of deepwater drilling projects.

National investigations revealed that the Macondo well blowout was anything but natural. Still, it involved natural elements—fire, water, the sea floor, and the air that fanned the flames. Soon after the catastrophe, BP's CEO, Tony Hayward, appeared on the *Today Show* to reassure the public in the wake of what he described as a "natural disaster." Perhaps this was a mere slip of the tongue from a man who quickly demonstrated a knack for sabotaging his own public relations efforts by saying the wrong thing at the wrong time. But one month later, a group of congressional leaders repeated Hayward's error when they asserted, "The oil spill in the Gulf is this nation's largest natural disaster and stopping the leak and cleaning up the region is our top priority."[13] If the Deepwater Horizon incident does not qualify as an *"unnatural disaster,"* it's hard to imagine what would.

* * *

THIS DISTINCTION IS particularly vexing—and important—in the Mississippi basin. In truth, it's a stretch today to describe the river as anything close to "natural." As the next chapter explains, over the past century, we have straightened, channelized, dammed, and rip-rapped the Mississippi, all in an effort to keep it in its channel and away from its natural floodplain. Ever higher levees and stronger floodwalls—supplemented by federal insurance and disaster relief—lure more and more people into the floodplain, which in turn requires more extreme measures to protect those people. Meanwhile, we have cut off the river's natural pressure-relief valve—its floodplain—leaving it no good place to go when the rains come.

Through these efforts, humans have demonstrated an uncanny ability to exacerbate the damage caused by natural hazards. Through our laws and public policies, through shortsighted and misguided engineering projects, and through poor individual decisions, we have increased, rather than decreased, the destructive power unleashed by natural forces. This book pays particular attention to the legal dimension of unnatural disaster. It tells the story of landmark laws and judicial decisions affecting flooding in the Mississippi basin, and focuses on three main topics. First, it traces how perceptions of the boundary between "natural" and "unnatural" disaster have shifted over time, affecting how the courts have assigned legal accountability for storm and hurricane damage. Second, it recounts Congress' response to roughly a century of flooding, telling the tales of seven of the most devastating floods in the basin's history, and Congress' modification of one policy after another, as seemingly good ideas revealed themselves to be recipes for disaster. Finally, the book concludes with specific suggestions for rethinking disaster responsibility and prevention, with an eye toward the fair and sustainable allocation of risk and the avoidance of so-called *moral hazards*—poorly conceived laws and policies that provide a soft landing for those who make unnecessarily risky decisions.

The stories of this book offer lessons to guide us in the future. They show humans at their worst and their best—prideful and humbled; arrogant and accomplished; foolish and industrious. The following pages contain tales of engineers, bootleggers, church camps, restaurants, and dynamite-wielding convicts, among others. Often, with the best of intentions, human actions have set the stage for unnatural disaster.

1

AN UNNATURAL RIVER

HOW WE GOT HERE

The Mississippi River, although mighty and impossing, has been bent to the will of humans. Throughout history, waterways have been magnets for civilization, and the Mississippi is no exception. But where some earlier societies adapted themselves to the rhythms of the river, the United States did just the opposite. Armed with the technology of the Industrial Revolution, American engineers of the nineteenth century began to remake the river itself in the name of trade, travel, and settlement. The changes have been wondrous. The engineers made the river straighter, shorter, and deeper. They redirected millions of tons of sediments away from floodplains, sometimes sending them far downstream into the deep recesses of the Gulf of Mexico. They lined the river with a wall of levees as far as the eye can see. They thought they could make the river predictable and unchanging. Our engineers have been re-working the Mississippi for such a long time that it is easy to stop seeing the changes. But taken together, the alterations have been so profound that the Mississippi River of today is far from "natural." When people settle near highly engineered river systems like the Mississippi—encouraged by legal policies and incentives—they sometimes forget the river's natural proclivity for change.

For more than one million years—long before engineers came on the scene—the area that would become the headwaters of the Mississippi River lay frozen under great sheets of ice. This frigid blanket entombed vast swaths of present-day Canada and the northern United States. In many places, the ice was over a mile thick. Grinding south into modern Wisconsin, glaciers captured trillions of tons of soil and rock. Their weight leveled the land, pulverized bedrock into powder, and depressed the earth's surface. The glaciers left behind debris in landforms still visible today, including *moraines* (mounds of rock and soil) and *drumlins* (low, elongated hills). Geologists tell us that the most recent ice age —the Wisconsinan—ended with a warming period a "mere" twelve to fourteen thousand years ago.[1] As glaciers melted and retreated, the groaning earth rebounded gradually—regaining up to 160 feet of elevation in some areas. Ice melt gouged the earth, unleashed torrents of water, and sculpted the landscape, leaving as its legacy the upper Mississippi River valley. As the icy symphony roared, the fauna of one era gave way to that of another. The five-ton mastodon plodded toward extinction. The saber-toothed cat unleashed its last snarl, baring its seven-inch fangs to the accompaniment of glacial water music.

The Mississippi's channel had barely formed when it began to change. The great river and its tributaries drain about 40 percent of the continental United States—well over a million square miles. As the river churns downstream, it grinds against its bed and banks with the accumulated force of that runoff. The river constantly remakes itself through three geologic processes: erosion, transport, and deposition.[2] First, the turbulent river erodes its bed and banks. The captured sediments dissolve or remain suspended in the water, contributing to the Mississippi's scouring power. Next, the river system transports an enormous sediment load—an estimated 180 million tons each year.[3] According to one calculation, it would take seventy-three million dump trucks to carry this annual load, enough to circle the earth twice, parked bumper to bumper. Third, rivers deposit their sediment load. When in flood, the Mississippi spreads out laterally and deposits clay, silt, sand, and gravel to form a broad alluvial valley. The coarsest (and heaviest)

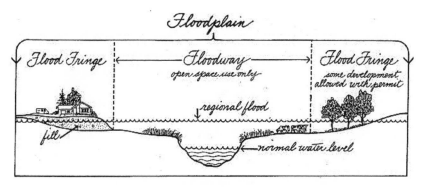

Floodplain protection standards: What is a floodplain?
Source: Wisconsin Department of Natural Resources

sands and gravels settle out first, typically mounding into natural *levees*, or embankments of earth, that parallel the stream channel. The finer alluvial silts and clays end up farther from the river. Historically, the sediments were so plentiful that the Mississippi created a floodplain almost one hundred miles wide in its lower section downstream of Cairo, Illinois.[4] This rich topsoil in the nation's heartland averages 132 feet in depth, enough to bury a thirteen-story building.[5]

As the Mississippi approaches the Gulf of Mexico, it deposits its remaining load. Over the past five thousand to six thousand years, sediment deposition has built up the current Mississippi delta, creating new land that extends some fifty miles into the gulf. Created from a patchwork of sediments gouged out of the thirty states in the river's floodplain, Louisiana's coastal plain protrudes out well beyond its neighboring state, Mississippi. Despite the seeming abundance of sediments, every soil particle is precious. Communities down in the delta depend on sediments flushed thousands of miles downstream for their livelihoods, and indeed their very lives. A vital building block of a town near the gulf today may have been, at one time, a particle of soil in Montana, Minnesota, or the Dakotas. These particles form on a time scale almost beyond human comprehension: it takes five hundred to a thousand years to create a layer of topsoil only one inch deep.[6] At that rate, the planet's newest inch of soil began to form as long ago as AD 1000, building slowly during the flourishing of the Incan empire, Hernando de Soto's explorations of the lower Mississippi River, and up to the present time.

Loving Water: A Fatal Attraction?

THROUGHOUT HUMAN HISTORY, waterbodies—especially rivers—have had an irresistible pull, promising prosperity, communication, peace, and even freedom. The Mississippi, in particular, has had a profound impact on American culture, history, and politics. Most Americans are familiar with Mark Twain's expositions on the river—through the eyes of Huck Finn or through Twain's autobiographical works—but fewer have seen the river in the way that one mixed-race descendent of slaves, slave-traders, and Cherokee Indians did. In 1921, poet Langston Hughes published "The Negro Speaks of Rivers":

> *I've known rivers ancient as the world*
> *And older than the flow of human blood in human veins.*
> *My soul has grown deep like the rivers.*
> *I bathed in the Euphrates when dawns were young. . . .*
> *I heard the singing of the Mississippi when Abe Lincoln went down to*
> *   New Orleans, and I've seen its muddy bosom turn all golden in the*
> *   sunset.*[7]

Perhaps Hughes' reference to Lincoln and the "singing" Mississippi was a reflection on the role of the river only a generation or two earlier, when crossing the Mississippi River into the state of Illinois, or crossing the Ohio River into Indiana or Ohio, was the most significant moment of a slave's journey to freedom via the Underground Railroad.

Whatever the inspiration, Hughes' poem is but one expression of the allure of the river. As a young steamboat pilot in the 1850s, Mark Twain found the Mississippi more intriguing than anything he had ever encountered in his young life—a beautiful and wise but tempestuous master. As he explained in *Life on the Mississippi*, "The face of the water became a wonderful book . . . delivering its most cherished secrets as clearly as if it uttered them with a voice. And . . . it had a new story to tell every day."[8]

Whether it is to the Mississippi River or to the most humble inlets, streams, and ponds, humans are drawn to water. Communities tend to situate themselves as close as possible to waterbodies.[9] Coastal areas make up only 17 percent of the landmass of the contiguous United States, yet more than half of all Americans live on or near the coast. They crowd in at a density of about three hundred people per square mile—three times more than the national average for non-coastal areas. And the density is expected to increase, with about 3,600 people relocating to the coast each day. The National Oceanic and Atmospheric Administration (NOAA) keeps close tabs on this trend. It reports that between 1970 and 2010, the population of coastal watersheds increased by 45 percent, placing millions more people in the path of future hurricanes and floods.[10] By 2025, 75 percent of Americans are expected to live in coastal counties.[11]

Are Americans more cavalier about the danger than others? Not really. To some extent, our modern societies are simply following in the footsteps of the ancients.

The cradles of civilization originated in river valleys and along the seashore. The Sumerians were based along the Euphrates and Tigris rivers. Egyptians built their empires along the Nile. The Indian civilization took root in the Indus River valley. The prosperous Shang and Zhuo dynasties centered on the Yellow River and the North China Plain. The Romans and Greeks occupied the shores of the Mediterranean Sea. These and other societies depended heavily on rivers, bays, and oceans to transport people and goods. They relied on fresh water from rivers for irrigating crops and for domestic needs such as drinking, cooking, and bathing.[12]

After all, water is elemental. Author Barbara Kingsolver described water as "the briny broth of our origins; the pounding circulatory system of the world."[13] Up to 75 percent of the human body is water. We can go for weeks without food, but if we try going without water we'll survive only a few days.

Although water is essential to human life, sometimes we forget that water is also dangerous. Despite the risks of storms, hurricanes, and floods, societies throughout history have gravitated toward water. But they soon learned to accommodate its mercurial nature, or suffer the consequences.

States Adrift

AS THE MISSISSIPPI snakes its way to the gulf, it carves a series of graceful, ever-changing curves, known as *oxbows*. Centrifugal force draws the scouring power of the Mississippi toward the outside of its curves, where it erodes the river bank. The loosened soil is either deposited against the inside curve of the next oxbow or transported farther downstream. Over time, as the river cuts through its oxbows, it follows a shorter, straighter, steeper route to the Gulf of Mexico. In some cases, the river creates these new short-cuts at a leisurely pace, through a process known as *accretion*. In other cases, the river can shift course abruptly, through *avulsion*. As described by Mark Twain in *Life on the Mississippi*,

> The Mississippi is remarkable . . . [in] its disposition to make prodigious jumps by cutting through narrow necks of land. . . . More than once it has shortened itself thirty miles at a single jump! These cut-offs have curious effects: they have thrown several river towns into the rural districts, and built up sand bars and forests in front of them.[14]

Despite the river's propensity for change, numerous states have established as their legal boundary "a line drawn along the middle of the Mississippi River."[15] But when the Mississippi shifts course, it can

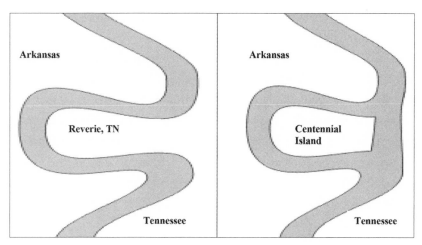

Before and after the Mississippi River shift of 1876
Diagram by Christine A. Klein

wreak havoc on the states that have unwisely relied on the constancy of the river. Arkansas and Tennessee learned this lesson the hard way, late one night in 1876.

At that time, a little more than thirty miles above Memphis, there was a great bend in the Mississippi known as Devil's Elbow. Reverie, Tennessee, was nestled in this crook of Satan's arm, on the *east* bank of the Mississippi. But on the night of Tuesday, March 7, 1876, the soil at the neck of the oxbow caved away. The river shifted course violently and gouged out a new shortcut right through the middle of its meandering channel. When the residents of Reverie awoke on Wednesday morning, they found themselves unexpectedly on the *west* bank of the Mississippi's new main channel. Overnight, they became island dwellers, stranded between the dwindling old channel to the west and the river's new primary channel now to the east.

In a matter of hours, the Mississippi swept away two thousand acres of farmland topsoil. Over the next few days, the river settled on its new route with uncontrollable force. Submerged lands were exposed, and dry lands were inundated. The river cut off nearly twenty miles of its length. Families barely escaped as the land beneath farmhouses, cotton gins, and barns crumbled into the newly formed riverbed.

The island, which became known as Centennial, found itself cut off from the rest of Tennessee by a new channel forty feet deep, uncrossed by roads or bridges.[16] Afterward, no one knew whether Centennial Island was part of Tennessee or part of its western neighbor, Arkansas. It took almost a century of legal wrangling and six trips to the U.S. Supreme Court to clear things up.[17] Eventually, in 1970, the Court decided that Centennial Island remained with Tennessee. When a river changes abruptly through "avulsion" as the Mississippi did in 1876, the Court explained, the pre-existing owner (Tennessee) gets to keep title. Conversely, if the river had changed gradually, through erosion or "accretion" of soils, the pre-existing owner would have lost title.[18] It all goes back to stability and settled expectations—what state would expect to lose something as important as title to its territory overnight?

Before these decisions, while court proceedings continued from the 1870s until 1970, Centennial Island was a sort of no man's land. This was particularly vexing to law enforcement officials in Arkansas and Tennessee, uncertain of who had jurisdiction over the island. And scofflaws like Andy Crum came perilously close to taking advantage of the legal void.

Andy Crum liked liquor. That's not necessarily a bad thing, except that he happened to live in a "dry" state at the dawn of the national Prohibition era. But being both entrepreneurial and fearless, Crum set up a bootlegging operation on Centennial Island in the early 1900s. Crum, it turned out, was not just a bootlegger. He was also a cold-blooded killer. When Arkansas Sheriff Sam Mauldin led an early morning raid in July 1915, the outlaw was waiting for him and shot the sheriff dead. Crum slipped away, but later that day the Arkansas deputies found him hiding in a cotton field on Centennial Island and locked him up in an Arkansas jail. Tennessee, eager for its own retribution against Crum for his many years of illicit endeavors, demanded that the prisoner be released into its custody. After all, Centennial Island occupied territory that originally belonged to Tennessee, at least back in 1876 before the river's shift.

As Tennessee and Arkansas wrangled over legal jurisdiction to incarcerate and prosecute Crum, local citizens took the law into their own hands. An angry mob was intent on vindicating the death of an officer of the law—whether or not Sheriff Mauldin had official jurisdiction over Centennial Island on the day of his death. The mob stormed the jail and murdered Crum as he sat in his cell. Afterward, Arkansas and Tennessee continued to fight over the ownership of Centennial Island. But the two states did agree on one thing: Crum's murderers would not be prosecuted. Frontier justice prevailed, and neither state brought charges against the vigilantes.[19]

Arkansas and Tennessee were not the only states to dispute boundary lines delineated by the Mississippi. As the great river adjusted its course over time—painstakingly carving out an oxbow here, taking a short-cut there—dozens of lawsuits followed. Neighboring states and private landowners all along the Mississippi River, from Minnesota down to Louisiana, sought judicial guidance to sort out their respective property claims.

Indeed, a careful look at a modern map bears witness to the river's propensity for abrupt change through avulsion. In general, the Mississippi River forms the western boundary of Wisconsin, Illinois, Kentucky, Tennessee, and Mississippi; likewise, it generally marks the eastern border of Minnesota, Iowa, Missouri, Arkansas, and Louisiana. But on closer inspection, one can pick out numerous sections where state boundaries continue to follow the Mississippi's long-abandoned route, rather than the current main channel. As a result, some states today straddle the Mississippi, with orphaned pockets of land stranded across the river from the rest of the state. These marooned areas mark the sites of abrupt river shifts, which sometimes occurred over the course of a single night.

The legal questions are vexing, but perhaps the bigger mystery is this: Why did humans attempt to set permanent boundary lines along the course of an ever-changing river in the first place? When engineers tamed the Mississippi system, they aimed for stability and predictability, two important pillars of prosperous communities and, not incidentally, of the American legal system as well.

Rivers and their soils, by nature, are dynamic. These two opposing forces—humans' need for stability and rivers' need for change—collide time and again.

Conquest

DURING THE EIGHTEENTH and most of the nineteenth centuries, humans adapted their watercraft to the sinuous, shallow channels of the Mississippi River and its tributaries. Following the example of the Native Americans, explorers and fur traders used shallow-bottomed, wooden boats for transportation and for moving freight. In 1705, the first documented non-Indian cargo—a load of fifteen thousand bear and deer hides—was shipped in canoes, or *bateaux*, by French *voyageurs* (travelers) from the Miami Indian country on the Ohio River downstream to the Mississippi River. For over a century, Europeans and their descendants continued using canoes, keelboats, pirogues, and other types of human-powered boats to move animal fur, especially beaver pelts, lead, and other products to and from the Mississippi River basin.

Commercial steamboating began in 1807 in the East with Robert Fulton's successful trip on the Hudson River, running from New York City to Albany. A few years later, Fulton and his partners built the *New Orleans* at a cost of around $40,000. She measured 116 feet long, weighed 371 tons, and had a paddlewheel mounted on her side. Nicholas Roosevelt, an ancestor of Presidents Theodore and Franklin Roosevelt, piloted her from the Pittsburgh shipyards to her namesake city via the Ohio and Mississippi rivers. She arrived in New Orleans in 1812, and was placed in service between that "Crescent City" and Natchez, Mississippi.

The *New Orleans*—following the model of Fulton's earlier steamboat—had been designed for deep, snag-free eastern rivers, and was no match for the Mississippi. During the War of 1812, Captain Henry Shreve worked out the structural and mechanical modifications necessary for use in shallow inland rivers. Shreve's design utilized a flatter bottom and a lighter engine, and added the popular upper passenger deck and pilothouse, giving Mississippi steamboats their distinctive "wedding cake" appearance.[20]

Shreve launched his innovative design with the *Washington*, a steamer built in 1816. With its "handsome private rooms and commodious barroom," as described by one historian, the *Washington* quickly became the industry standard for luxury travel.[21] When it steamed from New Orleans to Louisville, Kentucky—an important port linking the Ohio River and ports in the East to the Mississippi and ports in the West—the *Washington* secured the future of navigation on the Mississippi. The trip took about three weeks and broke all records for travel in that segment of the river. The St. Louis *Republican* showered Shreve with accolades for proving that the Mississippi could be navigated by steamboats. Just a few years later, Shreve's *Post Boy* became the first steamboat to deliver mail on a so-called "Western" river. For those times, the *Post Boy* was extraordinarily fast. When the U.S. Post Office Department commissioned it to carry letters between New Orleans and Louisville, delivery times dropped from a month to only a week, bringing residents of the far-flung regions of the nation much closer together.

When Shreve was not on the river, he was in court fighting Fulton's monopoly over government contracts for shipping on the lower Mississippi. Fulton and his colleagues had received exclusive steamboat licenses for navigating the rivers of both the state of New York and the territory of Louisiana. At one point, Fulton's lawyer had one of Shreve's steamboats seized and impounded, and subsequently he had Shreve himself arrested. But Shreve had no intention of giving up his interest in the Mississippi River, and he resumed control of his vessel as soon as his attorney could post bond.[22]

Shreve was not the only riverboat pilot to fight the Fulton monopolies, and in 1824 the issues came before the U.S. Supreme Court in the famous case of *Gibbons v. Ogden*. Aaron Ogden ran a ferry service from New York City to Elizabethtown, New Jersey, under a state grant that he had received from the Fulton syndicate. Thomas Gibbons, a former partner of Ogden, struck out on his own and began to operate a competing ferry service. Gibbons was crafty enough to obtain a license from the federal government under the Federal Coasting Act of 1793, thereby setting the stage for a battle between state and federal licensing authorities. Ogden obtained a legal injunction against Gibbons in state court, requiring Gibbons to cease operations, on the theory that navigation was a legitimate area of state, not federal, regulation. (Ironically, Ogden had served in the U.S. Senate from 1801 to 1803 as a Federalist, the party that was founded by Alexander Hamilton and that stood for a strong *federal* government.) Gibbons challenged the injunction, taking his cause all the way up to the U.S. Supreme Court.[23]

With Daniel Webster arguing his case, Gibbons was confident of success, and indeed, after hearing arguments over the course of five days, the Court ruled in his favor. (Litigants before the Supreme Court these days are given only a half-hour to make their arguments.) Although the notion of federal power over waterways was, at the time, contrary to the prevailing sentiment favoring local governance, the Court concluded that the Commerce Clause of the U.S. Constitution, which lodges power in the federal government to regulate "commerce . . . among the several states," covers more than just the trading of commodities; it also extends to "intercourse," including navigation. Chief Justice John Marshall found that the New York state grant conflicted with the federal statute and ruled that Ogden's interest must therefore give way to Gibbon's federally sanctioned license: "the State laws must yield to the superior authority of the United States."[24] Although the decision was relatively narrow in scope, it laid the foundation for expansive federal power over commerce and for the modern *preemption* doctrine, which makes federal law supreme to conflicting or inconsistent state laws.[25]

Since *Gibbons,* the Supreme Court has not hesitated to find state law preempted when it interferes with federal activities related to navigation, flood control, hydropower, or vessel safety.

Justice Marshall's opinion effectively ended the Fulton monopoly in the East and, just as Shreve had done on the Mississippi, opened up the steamboat lines and the nation's rivers to competition. By 1827, over one hundred steamboats plied the Mississippi River. By 1859, that figure had more than doubled, and the amount of cargo carried by steamboat to New Orleans increased twelve times over.

* * *

WHILE THE LAWYERS wrangled over steamboat monopolies and federal power over interstate commerce, the Corps of Engineers was out on the Mississippi River surveying its physical characteristics and navigational capabilities. Although earlier travelers had adapted their boats to the river, people now demanded that the river accommodate human demands. The invention of the steam engine and the construction of larger ships, in tandem with the settlement of the fertile river valleys by farmers and urban residents, prompted officials to seek a profound transformation of the river itself. In 1861, Captain A. A. Humphreys and Henry Abbot issued their renowned *Report Upon the Physics and Hydraulics of the Mississippi River; Upon the Protection of the Alluvial Region Against Overflow; and Upon the Deepening of the Mouths.*[26] This report established the Corps' *levees-only* policy, which continues to influence modern-day river management, especially on the lower Mississippi. The policy rests on the assumption that the construction of levees lining the banks of the river will constrain the water's flow and accelerate its current, which in turn will create sufficient force to scour the riverbed and deepen the river's channel, consequently enhancing navigation. In addition, at least in theory, a deepened channel can carry more water, thereby containing excess floodwaters in the river itself. Therefore, under the policy, levees alone should be sufficient for both navigation and flood control, without additional structures such as spillways and reservoirs to relieve the pressure of floodwaters.[27]

Navigational interests welcomed the Corps' survey and report. On the lower river, businessmen and politicians demanded action. Not only did they want the Corps to proceed apace with the construction of hundreds of miles of levees between Cairo and New Orleans; they also sought a sediment-free channel at the mouth of the river to support large vessels moving to and from the Gulf of Mexico. Engineer James Buchanan Eads—described by one historian as a man of "almost suicidal self-confidence"—was just the man to do it.[28] Fresh from his 1874 success in building a steel bridge across the Mississippi River at St. Louis—the first major steel bridge in the world, with the longest arch span ever when constructed—Eads was ready to tackle the gargantuan sandbars that were blocking ships at the river's mouth. Before that, the Corps had concocted an array of strategies to solve the problem but none had worked, and it had declared the sandbars a "permanent, immovable barrier."[29] Enter Eads, with a proposal for a system of huge *jetties*—devices much like levees, but projecting directly into the river rather than lining the river's banks. Eads hypothesized that, by constricting the Mississippi and directing the water toward the river's center into a narrow, fast-moving current, the jetties would force the river to cut its own deep channel through the sandbars all the way to the gulf.

The Corps, having tried modest jetties in this particular stretch of the river in an unsuccessful attempt to deepen the channel, opposed Eads' plan. Despite its "levees only" policy for the rest of the river, the Corps believed additional measures were necessary to promote navigation to the gulf and it proposed the construction of a deep-draft shipping canal instead. (No doubt, the Corps' obduracy also stemmed from the fact that its Captain Humphreys had become Eads' archenemy during the struggle for dominance over Mississippi River bridge-building.) After a great deal of lobbying and Eads' promise to pay for the jetty construction himself and to seek reimbursement only if his idea worked, Congress ultimately authorized Eads' plan. The authorization came with stringent conditions: the jetties had to be constructed through the smallest and shallowest pass—the nine-foot deep South Pass; they had to produce a thirty-foot deep channel; and Eads' reimbursement would be limited to $6 million. Originally, Eads had sought $10 million for construction through the longer, deeper Southwest Pass. It was a gamble, but Eads accepted Congress' terms and forged ahead with the job.[30]

By 1879, Eads' jetties had cut a thirty-foot deep shipping channel through the immense sandbars that had choked the mouth of the Mississippi for as long as humans had attempted to navigate it. The *New York Times—Daily Tribune* announced, "Genius, persistence and practical skills have seldom won so great a triumph over the forces of nature and the prejudices of men."[31]

Mississippi River trade doubled, and New Orleans rose to the second-largest port in the United States, trailing only New York. By then, Humphreys had resigned as Chief Engineer of the Corps, but even so, one might say that Humphreys had the last laugh. Just beyond Eads' jetties, the river soon began to deposit some of the sediments it had been carrying, creating a new sandbar where the South Pass met the Gulf of Mexico. Additional jetties were built and constant maintenance was required to keep the channel open.[32]

After the river's mouth was clear, St. Louis—situated strategically on the upper river near its confluence with both the Missouri and the Illinois rivers—wanted improvements to its own port. To enable ships to continue unimpeded upriver, Congress authorized the dredging of an eight-foot deep channel between St. Louis and Cairo, and subsequently approved dredging to nine feet. (The average *draft* of a European riverboat—the distance to the water's surface from the boat's lowest point—was six to nine feet; ocean-going vessels varied significantly but generally had drafts twice as deep.) With increased river traffic, cities sprang up in the river's floodplains, and ships and railroads competed for cargo at every port.[33] St. Louis became one of the world's greatest steamboat ports. Steamers delivered European immigrants from the eastern seaboard, ice from Minnesota, lead ore from Wisconsin and

Iowa, grain from just about anywhere in the basin, and, on occasion, theatrical troupes and even circuses with their menageries.[34]

* * *

IN 1882, AFTER many years away, Mark Twain returned to the Mississippi River. He climbed aboard the *Gold Dust*—as a passenger this time, not a pilot—with eager anticipation of reuniting with the river on a thirty-three day, three-thousand mile journey from St. Louis to New Orleans and then back north all the way to St. Paul. But it was not the same river that he had grown to love as a young man. Twain's 1882 voyage took place against the backdrop of tremendous change. The Civil War had come and gone, and America was open for business. Engineers, encouraged by the federal government, had been busy channelizing the river's meandering bends, inundating many of its perilous rapids, and dredging and deepening its shallow bed to allow ever larger ships to ply its waters.

The engineers liked to call these changes "navigational improvements." But to Twain, the river seemed "dead beyond resurrection." As he lamented in his journal:

> I didn't find Hat Island and upon inquiry learned that it has utterly disappeared. . . . I was able to recognize Grand Chain, but the rocks are all under water. . . . The river is so thoroughly changed that I can't bring it back to mind even when the changes have been pointed out to me. It is like a man pointing out to me a place in the sky where a cloud has been.[35]

Twain's consternation was more than that of an aging man nostalgic for the days of his youth. He could see that the Mississippi was rapidly becoming one of the most heavily modified river systems in the world. When Twain first traveled the river in the 1850s, he noted that it was the crookedest river imaginable, "since in one part of its journey it uses up 1,300 miles to cover the same ground that the crow would fly over in

675."[36] Within just a few decades, hundreds of river miles, along with the associated wetlands and their animal and plant communities, had been lost to channelization and straightening in the name of navigation and development.

* * *

UPSTREAM FROM ST. Louis, commercial navigation was slower to develop. Northern riverboat pilots were left more or less to their own devices on what the Corps described as the "notoriously unreliable Upper Mississippi River," which was dotted with rapids, shallow passages, and submerged hazards. In particular, St. Anthony Falls, the only significant waterfall anywhere on the river, prevented steamboats from reaching points above Minneapolis. Minnesota lumber companies got past this obstacle by floating their logs downstream, through the falls, to sawmills farther south. Other types of products could not be moved so easily.[37]

Farmers, mill owners, and cities in Minnesota, Iowa, and Illinois clamored for federal assistance. At the conclusion of the Civil War,

Locks and dams on the Upper Mississippi River

Source: U.S. Army Corps of Engineers

Congress authorized the Corps to create a four-foot deep navigation channel on the upper Mississippi. Major Gouverneur Kemble Warren, a West Point graduate noted for his leadership at the Battle of Gettysburg, was charged with surveying the river and its tributaries. Warren opened the first Corps district office in St. Paul, Minnesota, in 1866. He immediately acquired a dredge and snag boat for digging and maintaining the four-foot channel between St. Paul and St. Louis. Warren also authorized the construction of the first *wing dikes* on the upper Mississippi—a type of jetty constructed of rock and brush, extending from the shore partway into the river. Like other jetties, wing dikes force the water into a faster moving, deeper navigational channel in the center of the river.[38]

It soon became apparent that wing dikes and snag removal were inadequate to meet the region's growing commercial needs. By 1880, Minneapolis, St. Paul's Twin City, had become the "Flour Milling Capital of the World." At its peak, the Washburn A Mill was the largest in the world, grinding over one hundred boxcars of wheat per day to make enough flour for two million loaves of bread. (Today, the Minnesota Historical Society draws a parallel between the story of how flour milling at Washburn, which later became General Mills, propelled the region into the modern era and the story of how the Silicon Valley microchip changed the face of California one hundred years later.) Rail lines from the northern Great Plains brought grain to the Washburn A and other mills, while St. Anthony Falls supplied power for running the mills, and newly arrived Scandinavian immigrants provided labor. Yet the region still lacked reliable water-borne transportation below the falls to move the flour to the rest of the world.[39]

Initially, the flour millers of Minneapolis pressed the federal government to construct conventional dams to span the river and to impound water above the falls. They hoped that the steady release of water from the reservoirs pooled behind these dams would not only raise water levels for navigation, but also keep their mills turning longer and more consistently. In 1880, Congress authorized such a dam some two hundred miles upriver from Minneapolis, on Lake Winnibigoshish, a lake carved by glaciers ten thousand years ago.[40] At the time of the authorization, the region was inhabited almost exclusively by Chippewa Indians, who used the shores of the natural lake to pasture their horses and to harvest cranberries and wild rice. The dam, which was predicted to raise the lake's water level by fourteen feet, obliterated the Chippewa's homeland, rice fields, fisheries, and burial grounds.[41]

In order to gain ever greater control of the flows of the upper Mississippi River, Congress quickly authorized five more dams and reservoirs in the river's headwaters. In its subsequent reports, the Corps of Engineers boasted that releasing water accumulated in the six headwaters reservoirs had raised the water level in the Twin Cities of St. Paul and Minneapolis by a foot or more, much to the delight of both the navigation interests and the millers. Meanwhile, the St. Paul District of the Corps constructed over one hundred miles of wing dikes to better channelize the river and to maximize the benefits of the increased flows from the upstream dams.[42]

Farther downstream, another watery impediment had long thwarted flour millers and fur traders alike. Two long sets of rapids between Iowa and Illinois—the Des Moines Rapids and the Rock Island Rapids—blocked navigation between Keokuk, Iowa, and Moline, Illinois.

Even before it constructed the dams at Winnibigoshish and other headwater lakes, the Corps of Engineers tackled the Des Moines Rapids. An eleven-mile stretch through the Des Moines Rapids at Keokuk was completely impassable during low water. The Corps' initial attempt to blast a channel through the rapids was led by none other than Army engineer Robert E. Lee, who would later become the commander of the Confederate forces. His efforts on the river (and in the war, for that matter) were only marginally successful. After the Civil War, an eight-mile long canal was constructed to enable steamboats to bypass the rapids and to continue upstream from Keokuk. But even the canal was inadequate for the type and volume of river traffic occurring at the time, so Congress authorized a survey to determine whether the construction of a dam at the foot of the rapids would be more beneficial. The survey concluded that, by completely inundating the rapids, a dam would permit passage and cut travel time and operating expenses. In 1905, the Keokuk and Hamilton Water Power Company began to construct a forty-foot high dam for the dual purposes of promoting river transportation and generating electricity. Upon completion in 1914, the Keokuk Dam was part of one of the largest hydroelectric plants in the world. At the same time, the dam aided transportation by eliminating the Des Moines Rapids.[43]

Getting around the Des Moines Rapids would do a northbound captain little good, however, unless the Rock Island Rapids could also be bypassed. Early attempts to solve the problem involved the use of explosives to blow the rock out of the water in order to create a four-foot deep channel through the rapids. However, the currents were still fast and unpredictable, and the Rock Island Rapids remained a serious impediment to navigation until the completion of the Moline Lock in 1907. With the Moline Lock and the Keokuk Dam in place, steamboats and barges were able to run the river from New Orleans all the way up to Minneapolis. With trade came growth, and Minneapolis' population nearly doubled from 200,000 in 1900 to 380,000 in 1920.[44]

Even with steady flows through the four-foot navigation channel, larger ships had difficulty making it all the way up river. In 1907, Congress authorized deepening the navigational channel another two feet between Minneapolis and the mouth of the Missouri River at St. Louis.[45] The Corps was instructed to use whatever means necessary to create and to maintain this six-foot deep channel—blasting, dredging, constructing canals and locks, and anything else within its power.[46]

Despite the Corps' successes in deepening the channel and in removing natural obstructions, by World War I, navigation on the upper Mississippi was declining steeply due to the development of railroads, as well as to changing demands and distribution practices in milling. Virtually no through-traffic moved between St. Paul and St. Louis. Fearing that without a diverse transportation system, the Midwest would become an economic backwater, the region's business interests lobbied to revive navigation. This time, to accommodate larger, more powerful diesel-driven boats, a federal law enacted in 1930 directed the Corps to deepen the existing navigational channel *another* three feet, bringing it to a total depth of nine feet, which matched the channel depth below St. Louis.[47] To ensure constant water levels throughout the system, the law also required the Corps to replicate its previous success at the Moline Lock by constructing a total of twenty-nine locks and dams all the way from St. Louis up to St. Paul. As a result, rapids and falls, along with naturally occurring meanders, oxbows, and eddies, have been replaced with what the U.S. Geological Survey and the Corps call a "stairway of water."[48]

The Great Disconnect:
What's a Floodplain without a Flood?

TO ACCOMMODATE STEAMBOATS, tugboats, and barges, engineers dug successively deeper channels, incrementally covering more and more of the river's length. These efforts did much to enhance navigation and promote economic development. But in defiance of the "levees-only" theory, the alteration of the river would cause more frequent and more severe floods by constricting flows and driving water levels upward. As subsequent chapters will explain, engineers would venture beyond navigational improvements into the realm of flood control. As the increasingly navigable river attracted more residents, ever more attempts would be made to protect them from floods.

By herculean efforts, engineers wrested the Mississippi apart from its floodplain. They built floodwalls reinforced with steel and concrete. They compacted mounds of earth into levees to constrain the Mississippi's normal tendency to spill beyond its banks. The system of levees and floodwalls extends for more than two thousand miles along the lower Mississippi and its tributaries (measuring separately along each bank of the river).[49] According to some estimates, structures in the upper basin disconnect the Mississippi River from half of its floodplain.[50] In the river's lower section, the separation is even greater. Here, the river has been walled off from an estimated 90 percent of its floodplain.

The lure of settling in floodplains, seemingly protected by levees, has been irresistible. Farmers have cultivated the plains, enjoying the bounty of soils made fertile by a long history of periodic flooding. Riverside cities have developed into great centers of commerce and transportation. The relatively level terrain also provides a magnet for developers seeking easily buildable land at affordable prices. As a result, we have come close to creating that most unnatural of human constructs —the floodless floodplain. Those who settle there are gambling that the levees will hold.

After the nation started down the path of controlling the Mississippi River, it seemed impossible to stop. As Mark Twain said, engineers "have taken upon their shoulders the job of making the Mississippi over again—a job transcended in size only by the original job of creating it."[51] It is no mere boast when the U.S. Army Corps of Engineers states that "[n]o river has played a greater part in the development and expansion of America than the Mississippi."[52] But at what price? The following chapters take up that question, in the context of seven of the most severe floods in the basin's history.

2

A DECADE OF RECORD

FLOODS (1903–1913)

THE FEDERAL GOVERNMENT TACKLES

FLOODS, BUT WITH LEVEES ONLY

One of the earliest written testimonials of the Mississippi River's propensity to flood comes from Garcilaso de la Vega's chronicle of Hernando de Soto's expedition to the New World. In his journal, dated 1543, de la Vega described how the river's floodwaters rose up near its confluence with the Arkansas River, just across from modern-day Clarksdale, Mississippi:

> [A] mighty flood of the great river . . . came down with an enormous increase of water. . . . Afterward the water rose gradually to the top of the cliffs and [then] overflowed the fields with the greatest speed and volume. . . . On the 18ᵗʰ of March, . . . the river entered the gates of the little village of Aminoya in the wildness and fury of its flood, and two days later one could not pass through the streets of this town except in canoes. . . . And it was a most magnificent spectacle to behold.[1]

De la Vega reflected that, prior to the flood, an elderly Indian villager had warned the Spaniards that the Great River flooded every fourteen years. That particular year—1543—happened to be the fourteenth year. According to de la Vega's journal, his fellow conquistadors "scoffed at what the old woman said and cast it to the wind."[2] Yet they could see that the Aminoya Indians—who may have been related to Quapaw or Caddo people of the American Southeast—had built their settlements up in the hills to make way for the regular inundation. And because the Aminoya had learned to cultivate the rich soils of the floodplain, de Soto's expedition was able to satisfy its hunger with the corn, vegetables, and dried fruit that had been harvested and stored away, high and dry. Even so, somehow the expedition's members did not believe that the lessons learned by the Aminoya applied to them.

Archeological sites dotting the Mississippi River delta indicate that the native inhabitants of the Southeast were well adapted to their environment. According to anthropologist Tristram Kidder, they lived in small, relatively portable communities that could be moved to locations more favorable for hunting, fishing, and gathering.[3] Farther north, on the upper Mississippi, the Ojibwe and other Indian people also adapted their lifestyles to the river's recurrent floods. They harvested wild rice nourished by periodically saturated marshes; other food sources included plentiful fish, waterfowl, and muskrats and other river-loving mammals.[4]

But the Native Americans were by no means passive occupants. They left behind heaps of clamshells and fish bones, which accumulated high above water level in various places along the rivers and coasts. These mounds of domestic waste, known as *middens*, continue to impact the environment today, mostly in positive ways. They have become an important element of marsh ecology, hosting an array of unique plant species and exhibiting great biological diversity. Incidentally, the middens also provide high ground for floodplain residents to occupy in times of rising water.

In the years since de la Vega wrote his journal, historians and geographers have recorded dozens of floods on the Mississippi River and its tributaries, particularly the Missouri and the Ohio rivers. The elderly villager from Aminoya was not too far off when she informed de la Vega about the frequency of floods on the river. Since the early 1800s and the advent of regular recordkeeping, major floods have hit the Mississippi River basin every ten years or so.[5]

The early floods did little to discourage European farmers, who were drawn to the fertility of the Mississippi River's floodplain. Although they occasionally lost their corn and wheat to seasonal flooding, the bountiful yields of the next year's crops tended to make agriculture in the floodplain worthwhile. Southern farmers in the Mississippi delta liked to boast that "even a fencepost will sprout leaves," given half a chance.[6] Perhaps deep down they understood that seasonal floods are natural, life-giving occurrences on the Mississippi. Periodic flooding allows the river to flush sediments from its channels and to deposit rich soil in its oxbows, sandbars, and river banks. You could say that flooding is what makes the Mississippi River what it is; without floods, the river would be just a ditch, and its floodplains arid, lifeless land.

By the turn of the twentieth century, however, whatever lessons the delta farmers had learned about living in the floodplain had been lost. At the time, geologist W. J. McGee observed that "as population . . . increased, men have not only failed to devise means for suppressing or for escaping this evil [flood], but have with singular short-sightedness, rushed into its chosen paths. . . . [T]he flood remains . . . a hardly-appreciated obstacle to progress."[7] In McGee's mind, the problem was abundantly clear—increasing population density, occurring with the growth of cities in the river valleys: "the banks of the waterways bristle . . . with greater aggregations of population."[8]

The stories of these floodplain communities, and the government's initial efforts to satisfy them by conquering and controlling the river, are best portrayed by the floods of the early 1900s, which are the subject of the rest of this chapter. Some of the most notable events occurred on the Missouri River, the Mississippi's longest tributary.[9]

Kansas City Blues

THE FLOOD OF 1903 left an indelible mark on the new century. In terms of property damage and lost lives, it was the most devastating flood that residents along the Missouri River—in the mid-section of the Mississippi basin—had experienced since the beginning of European settlement.[10] Several large floods had occurred in previous years, particularly in the 1840s, but the Weather Bureau reported that, back then, the floodwaters ran "harmlessly over unbroken forests . . . tenanted only by the beasts of the field and the birds of the air."[11] (New Orleans was a notable exception. Its densely populated neighborhoods were especially hard hit in the flood of 1849, when a plantation levee broke upstream from the city, and thousands of New Orleanians were evacuated or perished.)[12]

By 1903, the landscape of the Midwestern states of Kansas, Missouri, Iowa, and Nebraska had changed significantly. When unusually heavy rains fell in May of that year, the Weather Bureau reported that the floodwaters that rose up from the Kansas and Missouri rivers found not forests, marshes, and pastures, but rather "highly cultivated fields, and . . . rich valleys filled to overflowing with vast industries devoted with never ceasing energy to the fulfillment of the insatiable demands of commerce."[13] Crops, railroads, stockyards, factories, streetcar lines, and entire cities had sprung up directly in the path of the flood.

On Memorial Day (which was also known as Decoration Day at that time), May 30, 1903, the *Kansas City Star* issued this account of the near-complete devastation of the Harlem neighborhood in Kansas City, Missouri:

> The water from the Missouri river is running through the streets of Harlem like a millrace. Every house in the town is flooded. All of the 600 inhabitants are practically homeless. . . . Crowds of spectators lined the south bank of the Missouri and many crossed the bridge to Harlem, where they watched the owners of submerged stables swimming their

horses and cattle to places of safety, and the removal of women and children from the houses in skiffs.[14]

The water jumped the banks of both the Kansas and the Missouri rivers, submerged Kansas City's central business district, and spread across the entire floodplain. According to the Kansas State Historical Society, in the West Bottoms, an industrial area where hundreds of poor immigrant families lived, "the rivers ceased to exist independently and instead formed a sort of inland sea."[15] Work stopped, and utilities like gas, potable water, and electricity went out. About twenty-two thousand people were displaced from their homes. Meanwhile, sixteen of the seventeen bridges spanning the lower Kansas River were washed downstream toward the Mississippi.

Sixty miles away in Topeka, where Kansas was just completing its capitol building, twelve feet of water submerged areas of the city. Twenty-four Topekans drowned. In all, the 1903 flood took approximately two hundred lives.[16]

Some Kansas residents blamed the railroads for exacerbating their flood losses. At the Kansas City depot, floodwaters overcame railcars filled with spring wheat. Unhappy purchasers of the wheat sued the Union Pacific Railroad for the loss of their grain, alleging that the railroad was negligent in failing to move the cars prior to the flood. The railroad argued that the wheat had been "destroyed by a flood of unprecedented character," and that the event should therefore be deemed "an act of God" for which it was not liable.[17] In resolving the dispute, a Missouri state court recited the general "black letter" rule of law that "where there is negligence concurring with the act of God, and but for such negligence the injury would not have occurred, the person guilty of the negligence will be liable." But in this particular case, according to the court, the railroad's failure to move the railcars was *not* negligent because the railroad had no "notice or expectation of such visitation of God"—a flood of this magnitude was not foreseeable. The court dismissed the case against the railroad under the following logic:

The immediate injury and result in this case was occasioned by the sudden great and unprecedented flood of 1903. It was a result almost altogether out of the course of nature. Its like had probably not occurred in the memory of any one living. Loss from such a cause was wholly unlooked for, and was not to be expected or even taken into consideration by the most cautious.[18]

Undeterred, the owners of a shipment of three thousand cattle initiated a separate lawsuit against the Atchison, Topeka, and Santa Fe Railway. When the 1903 flood washed out the ATSF's rail lines, the railroad took the cattle, which had been en route from Texas to South Dakota, out of the railcars and moved them to the Kansas City stockyards until the water receded. The livestock owners sought to hold the ATSF liable for the loss of over five hundred cattle that died of starvation or from severe injuries sustained during the flood.[19]

By 1903, the Kansas City stockyards were the second busiest in the nation, surpassed only by Chicago's. Over two million cattle and calves passed through Kansas City that year, and over 113,800 carloads transported livestock to and from the stockyards. The railroads and the stockyards had a great deal of expertise in the care and feeding of livestock, and even more expertise in keeping them moving.[20] However, on the night after the plaintiffs' cattle were moved to the stockyards, the Kansas River rose much higher than officials of the stockyards and the railroad had ever seen. To prevent the cattle from drowning, the ATSF drove them up onto overhead viaducts leading from one portion of the yards to another, where they remained for more than a week. A federal court in Kansas described how the cattle perished: "[O]n account of the flood conditions, the height of the water, the swiftness of the current, the floating of debris, the lack of equipment, they were not, and could not be, properly cared for." As in the grain case against the Union Pacific, the federal court held that the flood, not the railroad, caused the loss of the cattle, and therefore the ATSF was not liable. As the federal court explained:

That these acts of the defendant company furnished the occasion for the loss is unquestionable, but . . . the direct and proximate cause of the loss and damage was the unprecedented and unexpected flood and attendant disaster that came wholly without anticipation on the part of the defendant. . . . It was a serious loss and great hardship upon plaintiffs, but, in my judgment, it must be borne by them, as like losses are borne by others with property unfortunately situated in this unhappy valley in that ill-fated time.[21]

The common law of tort liability, which places responsibility for a loss on those who breach their duty of reasonable care and cause foreseeable injury to another, failed to provide a remedy for these plaintiffs. In both railroad lawsuits, the courts believed that, while some seasonal flooding along the Missouri River and its tributaries could be expected, the 1903 flood was an entirely different, and unexpected, event. The courts seem to have given little thought to the possibility that putting rail lines and stockyards in close proximity to the river in the first place may have constituted something less than reasonable due diligence, and may have indeed resulted in foreseeable harm to people like the plaintiffs.

If these cases had been tried just a few decades later, the plaintiffs may have had a better shot. At the tail end of the industrial revolution, courts tended to be more favorable to economic development and less sympathetic to laborers and other individuals harmed during the course of that development.[22] By the 1920s and 1930s, legal concepts of risk and foreseeability had begun to evolve to better respond to the broad range of harms resulting from improvident industrial conduct. Courts eventually developed a formula to determine whether a defendant had acted unreasonably in the face of foreseeable risks: if the burden of taking precautions against foreseeable harm is less than the probability of harm multiplied by the gravity of the resulting harm, and the defendant fails to take those precautions, then the defendant is liable for any harm that does occur. In the grain and stockyard cases, if the railroads should have known (or foreseen) that the livestock or the grain was likely to be damaged by the railroads' conduct during the flood, and the livestock or grain was in fact damaged by the railroads' conduct, then the railroads would have been liable if the burden of taking precautions against the foreseeable consequences was less than the likelihood and magnitude of harm to the plaintiffs' livestock or grain.[23]

In the early 1900s, however, private lawsuits proved unequal to the task of protecting citizens and their property from flooding, or of assigning blame and assessing liability for flood damages. If the courts wouldn't provide relief, then perhaps the federal government would.

At first, the U.S. Army Corps of Engineers was reluctant to involve itself in flood control, which it feared would divert resources away from its primary mission—promoting navigation. Throughout the late nineteenth century and into the twentieth, Americans generally subscribed to the notion of "self-protection"—individuals were responsible for defending themselves from flooding.[24] Farmers and townsfolk took it upon themselves to form local levee districts and to construct modest earthen berms along the rivers to contain floodwaters. When their levees held during small, seasonal floods, the residents were emboldened to redouble their efforts to reclaim the floodplain from nature's forces, in the belief that they could keep the land dry enough to promote more reliable crop production and more extensive urban development.[25]

Given the resistance of the Corps—the nation's premier engineering body—to assume responsibility for flood control, it is not surprising that Congress stayed out of the flood control business for decades. In fact, in one of its early appropriations for navigational improvements, Congress stipulated that *no* federal money could be used to protect land along the Mississippi River from flooding or for any purpose other than navigation. While influenced by the Corps, Congress was also acting according to the prevailing nineteenth-century belief that, as a constitutional matter, federal powers over inland rivers were limited to commercial shipping and navigation. Although the U.S. Supreme Court had held unequivocally in *Gibbons v. Ogden* (the case that ended

Robert Fulton's steamboating monopoly in the East) that navigation was a matter committed to federal authority, it had not had occasion to address federal flood control.[26] Matters beyond navigation, many thought, were within the sole purview of state and local governments. Federally elected officials also feared the political and financial burden that could be expected if the federal government took on the challenge of nationwide flood control.[27]

Floodplain farmers and townsfolk saw things a little differently. For years, residents of Kansas and Missouri had sought federal navigational enhancements to improve the Missouri River's capacity for barge traffic, but after 1903 they began to push hard for federal flood protection as well. Local politicians and newspapers led the effort to attract attention —and money—to their cause. In 1907, Midwestern humorist George Fitch published a popular "character sketch" of the Missouri River, describing it as a river "that goes traveling sidewise, that interferes in politics, . . . cuts corners, runs around at night, lunches on levees, and swallows islands and small villages for dessert. . . . Blessed be the man who shall first find a way to chain it down and pull its teeth."[28]

That same year, the Kansas City District Office of the Corps of Engineers was created to oversee navigation on the lower Missouri River. The arrival of a second flood in 1908 added to the residents' sense of urgency in convincing the federal government to "chain [the river] down and pull its teeth." Although the 1908 flood was not as voluminous as that of 1903, farmers suffered even greater losses because flooding occurred just before the wheat harvest and destroyed the crop. Congress responded by providing a regular stream of funding for channelization of the Missouri River from Kansas City to St. Louis, but for navigational purposes—*not* flood control.[29] The appropriations placated Missouri River valley residents because they believed that dredging a deeper navigational channel and building wing dikes and stone revetments along the river's banks could accomplish both purposes.

Not to be outdone, residents of the Mississippi River valley and along the Mississippi's other major tributary, the Ohio River, would soon demand attention from Congress and the Corps when they, too, experienced successive record-breaking floods in 1912 and 1913.

Up on the Rooftops

IN FEBRUARY 1912, extraordinary amounts of snow fell over the Missouri and Mississippi rivers, blanketing Missouri and Iowa with drifts of white. Then, instead of bringing the first hint of spring to Great Plains residents pummeled by the long, harsh winter, March delivered the greatest monthly snowfall to Nebraska and Kansas in at least thirty years. Meanwhile, a series of six storms in the East dumped heavy snow and rain on the Ohio and two of its tributaries, the Cumberland and Tennessee rivers. Memphis, Tennessee, experienced 9.53 inches of rain in March, compared to its normal 5.77 inches. The height at which flowing water threatens lives or property is known as *flood stage*. By the end of March, the water level at Memphis was three feet above flood stage. In April, as the floodwaters moved downstream toward the lower Mississippi River basin, rain began to fall in southern Illinois, Kentucky, Arkansas, Missouri, Mississippi, and Louisiana, with rainfall in the lower basin ranging from four to eight inches above normal.[30]

By 1912, the Corps had built a series of dikes and levees along the lower Mississippi River to promote commercial navigation. (As noted in chapter 1, several strategically located sets of locks and dams on the upper Mississippi were completed by this time as well, to support commerce upstream.) In addition, local levee districts throughout the lower basin had constructed relatively modest earthen embankments to protect plantations and floodplain residents from submersion by high water.[31]

The levees could not withstand the strain of one million cubic feet per second of water—at least three times the average flow—barreling down the river channel. The first reported *crevasse* (that is, a breach, rupture, or break in a levee) occurred near Hickman, Kentucky. It took only a few hours for the floodwaters to reach the rooftops of houses.

"All I Saved Was the Bible": Photograph by J. C. Coover of refugees
on the levee, 1912 flood

Source: Memphis/Shelby County Library Collection, Memphis, TN, Memphis No. 1457

Eventually, 310 square miles situated behind the Hickman levee were
submerged. Although the crevasse released some of the pressure on the
river, floodwaters continued to rush downstream.

When they reached Caruthersville, Missouri, one thousand men—
half of the town's population—joined together to place sandbags at the
top of the railroad embankment that shielded the town from the Mississippi River. Despite the efforts of this army of flood fighters, thousands
of feet of the embankment crumbled. Crops throughout the region were
destroyed, and livestock losses numbered in the thousands. In Mississippi County, Missouri, an agricultural area in the southeastern corner
of the state, surrounded on three sides by the Mississippi River, a few
families were able to save their livestock by stabling them on top of an

Indian mound (a burial site or perhaps a midden heap), but many lost all of their horses, cattle, and hogs.[32] Witnesses described the stench of bloated and rotting carcasses caught in treetops and washed up against fences and buildings as unbearable.

Farther downstream in Wilson, Arkansas, Robert E. Lee "Boss" Wilson, a shrewd businessman and the town's namesake, watched the floodwaters rise from the second floor of his own hotel. Boss Wilson had founded Wilson as a company town for his sawmill and logging crews in 1886, and if he went under, the town would surely follow. The 1912 flood took his lumberyard, much of his Jonesboro, Lake City & Eastern Railroad, and his cotton crop. But Boss was just the kind of man, with just the requisite amount of determination, to rebuild. Although he retreated, temporarily, to his unflooded plantation over in Armorel, Arkansas, Boss quickly began laying plans to secure financing, supplies, and labor to rebuild the town of Wilson.[33]

Next, the floodwaters reached Greenville, Mississippi, where the 1912 flood revealed an ugly side of America all too common in the post-Reconstruction era. There, backed up by Ku Klux Klan nightriders and lynch mobs, Jim Crow laws flourished. Blacks were routinely subjected to economic servitude by wealthy planters who acted as if slavery were still perfectly legal. During the 1912 flood, an enterprising young engineer who ran out of sandbags near Greenville "had a brilliant idea," according to the *New York Times*. He ordered several hundred black convicts to lie down on top of the levee while water rushed over their bodies, thus creating a "human dike" to hold back the water. After all, the blacks were "standing idle" anyway, and the engineer thought they may as well be useful. The *Times* reported:

> Negroes Save Day: For an hour and a half . . . the negroes uncomplainingly [stuck] to their posts, until additional sandbags arrived. Then the human wall was replaced and the capping was pronounced sufficiently strong. The feat was somewhat similar to that performed at Baton Rouge, La., in 1852, when a levee broke [and] three thousand negro

slaves . . . were ordered into this breach, and held back the waters for twenty-four hours, saving the levee and preventing a heavy loss of life.[34]

As the floodwaters continued downstream through Louisiana during the first week of May 1912, the people residing behind the levees in Simmesport, Vicksburg, and Baton Rouge were in a state of panic. A newspaper reporter later reminisced about his grandfather's house, near the confluence of the Atchafalaya and Mississippi rivers: "Grandpa Lindsey's house was on stilts so high that I could stand under the house as a child. But the 1912 flood put 3 feet of water in the elevated first floor. . . . At the height of the flood, [mother] said she saw snakes swimming down the first floor hallway."[35]

Down in New Orleans, as reported by the *New York Times*, "every available man is at work, and lawyers, doctors, and business men are handling spades and picks side by side with negro roustabouts and convicts" in an attempt to fortify the levees.[36] The efforts of this unusual work crew were no match for the Mississippi River. On the night of May 10, the river rose eight inches in only two hours. By 10:00 p.m., the river gauge at Canal Street in New Orleans was at 21.9 feet, an inch or so higher than the maximum flood stage predicted by the Weather Bureau. High winds swept the waves over the levees, and the *New York Times* reported that "the streets of New Orleans were flooded as never before in the city's history."[37] The water was nearly seven inches deep on the sidewalks, and a strong current rushed through the streets of the French Quarter and up into the stores on Canal Street. Assessing the catastrophe, Louisiana Governor J. Y. Sanders reported that the flood had left 350,000 people homeless and caused $6 million in property damage in that state alone.[38]

In all, the floodwaters swept through twenty-two crevasses totaling around twenty-two miles in length, inundating nearly seven thousand square miles (about 30 percent of all of the floodplain land behind the levees). At seventeen of the eighteen gauging stations south of Cairo, the 1912 flood exceeded the highest readings ever recorded since the gauges were first established in 1871.[39] Although there appears to be no accurate count of the total number of human lives lost during the flood, at least several hundred died from drowning, disease, exposure, and starvation. By some accounts, three-fourths of those who drowned were black.[40]

* * *

REMARKABLY, THE NEWSPAPERS gave the 1912 flood relatively scant attention. Another water-borne disaster had stolen the headlines. The Virginia Newspaper Project, part of the state Library of Virginia, combed through the news archives and concluded that the loss of the supposedly unsinkable R.M.S. *Titanic* on April 15, 1912, sidelined every other piece of news for weeks. The researchers explained:

> What made the *Titanic* sinking so newsworthy was not simply the 1500+ death toll. . . . Rather, the *Titanic* became a legendary story for other reasons. The sinking of the *Titanic* tells a primal tale of man challenging nature and losing. Some disasters are undeniably acts of God, but the *Titanic* sinking has always seemed ambiguous. Although caused by an iceberg, it was also man-made, the result of the state of mind— grandiose, avaricious, and self-confident—of the British and American magnates and engineers who conceived and built the ship.[41]

The *Titanic* symbolized what the Newspaper Project described as "early 20th century industrial vigor, . . . a naive confidence in the boundless limits of technology in man's conquest over nature"[42] —not unlike the navigational channels, levees, and other engineered devices of the Mississippi River. There, by confining the river to an artificially narrow channel, the local flood control levees and the federal navigational structures (those that remained intact, anyway) produced higher flows than would have occurred naturally. Such faith in technology and engineering aggravated flooding problems downstream by increasing the magnitude of the floodwaters.[43]

In his analysis of the 1912 flood, published a few months after it occurred, geographer Robert Brown warned:

> Whenever a great flood passes down the Mississippi River, the possibility of a greater is at once apparent. And as the settlements on the flood plain are tending to increase, partly because of the protection during periods of minor floods given by the levees, . . . the problem grows yearly more and more intricate. . . . [U]nless the extreme volume of water which can be figured as a possibility in any Mississippi River flood can be cared for by the levee system alone, the public money should be turned to some more efficient means of protection.[44]

Brown demonstrated foresight in predicting an even greater flood in the future, but he probably didn't think it would come so soon. When the Mississippi River rose again in 1913, much of the damage sustained in 1912 had not yet been repaired. For floodplain residents living behind the compromised levees, the stage was set for disaster.

* * *

EARLY SNOW MELTS and heavy rains came to the Ohio River valley in the spring of 1913. The water began to rise, first on the Ohio and then on the Mississippi. Residents feared the worst. In some parts of Ohio, where the Ohio and Erie Canal and other navigational structures had been in place since the early 1800s, officials dynamited canal locks in an attempt to alleviate flooding by releasing pent-up flows that had risen behind the locks. Seven locks were blown up in Akron, allowing the floodwaters to pour from the canal into the Cuyahoga River and then into Lake Erie.

In Cincinnati, the Ohio River rose twenty-one feet in just one day. Levees along the Ohio River at Portsmouth were topped, flooding forty-five hundred homes. In the state capital, Columbus, a seventeen-foot wall of water from the Scioto River smashed through residential neighborhoods, forcing people to seek refuge anywhere they could

Vintage postcard of the 1913 flood
Source: Courtesy of the Dayton Metro Library, Dayton, Ohio

—including on rooftops and in trees. Thirteen people were rescued from a single tree.[45]

The most severe flooding occurred along the Great Miami River, a tributary of the Ohio River. When the Great Miami's levees gave way, water rose up to twenty feet in downtown Dayton. Sewage treatment at that time was rudimentary, and the flood swept away every outhouse in town, along with its contents. As gas lines ruptured under the force of the flood, fires broke out across the city, but the fire department was unable to get to the fires to put them out.

The 1913 flood made its mark throughout the Midwest. Newspapers carried accounts of storms and flooding all the way from Nebraska to Indiana. In Indiana, Easter Sunday, March 23, arrived with a downpour that continued for five days, inundating Indianapolis, Terre Haute, Muncie, and Peru. Thirteen inches of rain—the amount that normally occurs over a two-month period in this region—fell in just twenty-four hours.

As the 1913 floodwaters raced downstream, they gathered all the force that the Ohio River, the Great Miami River, and other Mississippi River tributaries could muster. The Memphis District of the Corps of Engineers, which extends 355 river miles from Cairo, Illinois, downstream to the mouth of the White River in Arkansas, once again prepared for the worst. And once again, Kentucky experienced the first of several levee crevasses. Laborers—many of whom were black men held at gunpoint—had been working for days to shore up the levee protecting Columbus, Kentucky, but on March 31, 1913, they abandoned the hopeless task. By 5:00 p.m. that day, the Columbus levee had been breached in several places, and in just a few hours, five to ten feet of water submerged the town. On April 1, a three-hundred-foot slab of the levee at Greenfields Landing, Missouri, just across the Mississippi River from Cairo, gave way. As in 1912, water quickly covered most of Mississippi County, Missouri.[46]

Nine days later and one hundred miles downstream, dozens of levee workers on the Mississippi River near Wilson, Arkansas—Boss Wilson's company town—abandoned the site in the face of an approaching storm. Strong winds drove the water over the sandbag topping they had

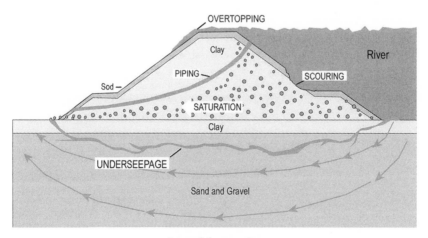

Levee failure modes
Drawing by Dr. J. David Rogers

been working on, and one hundred feet of earthen revetment collapsed "with a roar that could be heard for a mile or more," according to the *New York Times.*[47] Within minutes, a three-hundred-foot gap formed and water began pouring into Poinsett, Cross, and Crittenden counties in Arkansas, sweeping away forty-five men who had remained on the levee. The lower St. Francis River, which joins the Mississippi near Helena, Arkansas, overflowed, inundating over a half-million acres of northeastern Arkansas and southeastern Missouri.

On the east side of the Mississippi River, in Memphis, Tennessee, the flood water quickly found the most vulnerable spot in the patchwork levee—the brick wall of an old abandoned building belonging to the Stewart-Gwynne Cotton Company. On April 5, the wall collapsed, and other parts of the levee eventually went with it. Twenty city blocks were flooded.

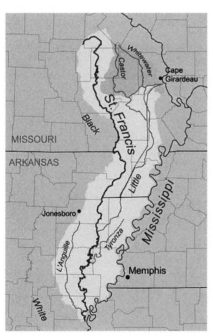

Map of the St. Francis River watershed
Map by Karl Musser (based on USGS data)

Most railroad operations in the area were halted due to washouts, with one significant exception. New Orleans prepared itself for the impending flood by ordering steel sheet pilings from the Carnegie mills in Pittsburgh to shore up the existing levees. Steel workers toiled around the clock at Carnegie's Homestead mill to meet the order, and the federal government secured an unobstructed right-of-way for the train that rushed south with the steel sheets.[48] Manufactured and shipped in record time, the pilings arrived in New Orleans on April 14, only a week after the order was placed. As the floodwaters entered the city, officials exuded confidence: "The levees are in splendid condition and can stand all the water predicted by the government . . . [and] there is not the slightest danger of a break in this vicinity."[49] This time, the optimism was well placed. The city experienced some flood damage, but nothing like the extensive losses felt upstream.

By the time the Mississippi's floodwaters had receded, the flood of 1913 had caused over six hundred deaths, mostly in Ohio. In comparison to the 1912 flood, the 1913 flood attracted a great deal of media attention. It made headlines throughout the country and spawned sensational postcards proclaiming "the greatest flood in world history."[50] But that dubious distinction would be short-lived as people continued to settle and build their residences and businesses in the floodplain, moving themselves directly into the path of the next great flood. It would make its appearance just fourteen years later.

* * *

WHEN MRS. ROBERTA G. WATTS' low-lying property flooded in 1913, she, like many litigants before her, blamed the railroads for her losses. Specifically, Mrs. Watts claimed that the floodwaters were trapped by the railroad trestle and embankments, which caused the Wabash River to back up and inundate her adjacent parcel. Her husband, who happened to be a civil engineer, had warned the railroad that the trestlework would not allow water from the Wabash River to disperse during a flood, but the railroad disregarded his opinion and went forward with construction. When the disagreement reached the Indiana Court of Appeals, the court agreed that human conduct had contributed to the damage, in striking contrast to the 1903 flood litigation that refused to hold railroads accountable for damages to the plaintiffs' livestock and grain. The Indiana court, validating the predictions of Mr. Watts, concluded that the trestlework and embankments created "foreign" conditions that channeled the water's natural flow into "torrents whose volumes, velocities, [and] force . . . operated as to scour, gouge out, and excavate . . . a hole in the ground 200 feet in width by 300 feet in length, and with a maximum depth below ground surface of 28 feet, and to cast, scatter and distribute a vast tonnage of earth, sand, fine gravel, and coarse gravel" upon Mrs. Watts' land.[51]

Mrs. Watts fared much better in court than had the grain and stock owners in the railroad lawsuits following the 1903 flood. Agreeing with the Court of Appeals, the Indiana Supreme Court found that the railroad's lack of due care in designing and constructing its embankments and trestles, and in failing to provide adequate openings to accommodate the flow of water, was "a direct and proximate cause which operated in conjunction with the act of God in producing the injury."[52] The court concluded that the company should have foreseen damage to low-lying property owners such as Mrs. Watts. It found the company liable for its negligence and ordered it to pay for the damages to Mrs. Watts' land.

Thus, some early twentieth-century courts recognized that at least some flood damage was an *unnatural* disaster. Would this lesson stick in the decades to come?

Who's the Boss?

FOR THE SECOND time in just one year, Boss Wilson lost his cotton crop, and his financial empire began to crumble under the weight of the consecutive floods. But Boss refused to give up on his fertile Arkansas land. After the floodwaters receded, he convinced a group of Midwestern investors to lend him $400,000 to help him get back on his feet, and before long he had amassed thousands of acres of cropland along the Mississippi River. Boss' experiences with the 1912 and 1913 floods did not deter him from developing the river bottoms; instead, he led the effort to form drainage districts in the county, and persuaded other landowners to help him drain the marshy ground. By the 1930s, Boss' enterprise—Lee Wilson & Company—operated one of the largest cotton plantations in the South. When he died in 1933, Boss Wilson left an estate that included sixty-five thousand acres of farmland as well as a Ford dealership and several other businesses and residences in and around Wilson, Arkansas. The town of Wilson continued to operate as a company town for many years. Until the 1940s, the only residents were Wilson Company employees; subsequently, the town incorporated and now has a population of about nine hundred. According to University of Arkansas professor Jeannie Whayne, the Wilson Company, managed by the fourth generation of Wilsons, continues to exercise "considerable economic and political influence" in the region.[53]

Although Boss Wilson was able to get back on his feet, the 1912–1913 floods left most Mississippi valley residents physically and economically exhausted. The local levee districts, in particular, were destitute. Frustrated floodplain landowners launched a massive propaganda campaign aimed at securing an unequivocal federal commitment to control future floods in the Mississippi River valley. Leading the charge was LeRoy Percy, a wealthy plantation owner from Greenville, Mississippi, situated about two hundred miles downstream from Boss' empire. Percy, who served as a United States senator from 1910 to 1913, had even more political clout than Boss Wilson, and at a much higher level.

In response to pressure from Percy and his allies, President Woodrow Wilson directed an entity known as the Mississippi River Commission to submit a comprehensive report on flood control. The commission had been created by Congress in 1879 to study Mississippi River flooding and the effects of the "broken chain of levees . . . theretofore constructed by state and local authorities."[54] It was led by the Corps of Engineers and included a representative from the U.S. Coast and Geodetic Survey (now part of the National Oceanic and Atmospheric Administration), as well as a few civilians. Although the commission was authorized to investigate and improve the conditions on the river, its power was limited, as the U.S. Supreme Court emphasized:

> Nothing justifies the conclusion that . . . Congress assumed control of the entire work of protection from overflow by levees, to the displacement of the state or local authorities. On the contrary, . . . the levees built

by the United States in aid of navigation at the same time afforded protection from overflow, and thus served a twofold purpose, that thereby
renewed energy was stimulated in state and local authorities to undertake the work of building levees for protection.[55]

The commission's report to President Wilson considered six methods of flood control. In addition to reforestation, the report evaluated
the effectiveness of constructing reservoirs, cut-offs, outlets, floodways,
and levees. Its principal author, Army Colonel Curtis Townsend, pre-
ferred the latter—the long-standing federal policy of "levees only,"
traceable back to the work of engineer James Buchanan Eads in
the 1870s. In Townsend's view, levees alone should be sufficient for
both navigation and flood control, without recourse to the other five
approaches considered in the report.[56]

The "levees-only" policy was not without its skeptics. After all,
hadn't the floods of 1912 and 1913 proved to be too much for the existing
levees? Opponents of the policy, who sought more holistic measures,
included former President Theodore Roosevelt (1901–1909), his chief
conservation officer, Gifford Pinchot, and Senator Francis Newlands of
Nevada, who was the architect of the Reclamation Act of 1902, which
authorized the construction of dozens of huge federal dams to supply
irrigation water throughout the West. Roosevelt and his fellow Progressives wanted nothing less than full-blown multiple-purpose river
development, complete with dams, reservoirs, manmade diversions and
outlets, and reforestation. But Congress was unpersuaded.[57]

Like the Progressives, Sir William Willcocks, described by the *New
York Times* as "one of the British Empire's most distinguished engineers," criticized the United States' levees-only approach to Mississippi
River flood control. Willcocks knew a thing or two about unruly rivers,
as he had designed the Aswan Dam on the Nile and had also played a
role in regulating the Tigris and Euphrates rivers in Mesopotamia. In
1914, during a lengthy interview with *Times* reporter Edward Marshall,
Willcocks explained his reasoning:

The general scheme of your endeavors to control the Mississippi seems to have been characterized from the start by the conviction that a river can be put into close confinement along its entire length straight-away. Such confinement you produce by building embankments, or levees as you term them. . . . [Y]ou have allowed the levees to be put up quite irregularly, and as each upper levee has been erected floods have worried those down below. . . . You have deprived your river floods of the opportunity of spreading out at the sides by contracting them, and so, perforce, they have risen into the air—risen frequently, until, with dire results, they have overflowed the natural and artificial barriers which normally confine them. . . . Therefore it seems to me that you really have brought the greater portion of your vast flood problem upon yourselves by bad management.[58]

One can imagine that Willcocks relished the opportunity to bash American engineering and, even more broadly, the American psyche. "The problem of the Mississippi is a fascinating one, but more a problem of your national psychology than of your river," he said. If people could continue to live with the Nile, the Tigris, and the Euphrates, then why couldn't Americans get along with the Mississippi? Willcocks taunted, "You treat the Mississippi as if it were a river apart, differing utterly from all other streams. It is nothing of the sort."[59]

According to Willcocks, three million acres of the St. Francis Basin, which was "nearly all primeval forest," should be set aside from development so that it could serve as a natural flood control reservoir. Even more galling, at least for state and local politicians, Willcocks suggested that the United States adopt "a concrete, comprehensive Central Governmental scheme, to which the States must bow." He added that under such a scheme, the federal government could exercise its navigational authority to prevent local authorities from constructing flood control levees in unsuitable places. His theory was that every human-wrought impediment to the flow of floodwaters "disorganizes the river and hurts navigation."[60]

Congress and the Corps ignored both Sir Willcocks and Roosevelt's Progressives. Colonel Townsend wasn't necessarily against additional measures in order to maintain navigation, generate power, and inhibit soil erosion, but he stood by the levees-only approach when it came to flood control. Townsend and other Corps officials reasoned that levees just needed to be higher and thicker to withstand whatever the Mississippi could throw at them. Besides, they argued, any other option would stray too far beyond the Corps' primary mission—navigation—and, worse yet, would cede too much control over navigable waters to a hodgepodge of local and state entities rather than the Corps. Equally troubling to politicians and landowners alike, reservoirs, outlets, and reforestation would result in a loss of developable land in the floodplain and would simply cost too much.

The outbreak of World War I in 1914 tipped the scales in favor of the Corps and against any type of comprehensive federal river management. As overseas trade revenues dropped, President Wilson withdrew his support for Senator Newlands' multiple-purpose development bill and called for deep cuts in federal spending. Navigational improvements remained high on the list of domestic priorities, however, as preparations for military conflict taxed the railroad network and water-borne shipping became the most attractive alternative for alleviating traffic congestion.

On March 1, 1917, just one month before it committed the United States to war against Germany, Congress enacted the Ransdell-Humphreys Flood Control Act. Named for its sponsors, Senator Joseph Ransdell of Louisiana and Representative Benjamin Humphreys of Mississippi, the act committed the federal government to the "levees-only" strategy and the federal treasury to expenditures for the construction of flood control levees in the Mississippi River valley. The act required new levees to meet federal construction standards and existing levees to be fortified and raised at least three feet above the high water mark of 1912.[61]

The act was a relatively limited measure in that it did not require reservoirs, outlets, or any other flood control measures, and it shied away from any hint of land-use restrictions. For the first time, however, the 1917 act authorized federal levee construction even if the levees created no direct benefit to navigational capacity.

By 1917, Congress had gained confidence in its legal authority over flood control. In two cases issued in 1916, the U.S. Supreme Court had alleviated the constitutional concerns about federal authority to engage in flood control activities. In *Cubbins v. The Mississippi River Commission,* the Court upheld the federal government's power to build levees along the Mississippi River to protect land from "extraordinary overflows," even though the construction exacerbated flood problems elsewhere on private land fronting the river near Memphis, Tennessee.[62] Similarly, in *U.S. v. Archer,* the Court did not question whether federal authorities had the constitutional power to construct levees and dikes on the lower Mississippi, but only whether the government must pay compensation to a plantation owner when the structures increased the height of flood waters channeled onto his land, which had the misfortune of being situated beyond the levees' protective zone. (Interestingly, the plantation owner himself "concede[d] the power of the government over the river.") The Court found that the owner had failed to prove that the United States was at fault, and remanded the case for further fact-finding. It explained: "Great problems confronted the national and state governments; great and uncertain natural forces were to be subdued or controlled, great disasters were to be averted, great benefits acquired. . . . [M]any forces were at work, and if the conditions at claimants' plantation were artificial, they were the result of the lawful exercise of power over navigable rivers."[63]

Under the 1917 Flood Control Act, project implementation was to be supervised by the Mississippi River Commission, which continued to be led by the Corps. The commission's jurisdiction was expanded to include the Ohio River and other watercourses connected with the Mississippi to the extent necessary to control floodwaters originating from any of those rivers. Although the breadth of the commission's jurisdiction reflected a dawning realization that engineering efforts on the upper river and its tributaries had consequences for the lower river, the commission's initial efforts were limited to the lower Mississippi. Congress allocated $45 million for levee construction between the mouth of the Ohio at Cairo and the mouth of the Mississippi below New Orleans. Congress was not willing to foot the entire bill for the work, however, and local entities were required to contribute one-third of the cost of levee construction and also to maintain the levees once they were completed.[64]

* * *

WHILE LEVEE CONSTRUCTION was proceeding, former Senator Percy, having obtained a measure of success with the passage of the 1917 Flood Control Act, found himself back in Mississippi cooling his heels. Percy's efforts on behalf of the state of Mississippi failed to earn him a second term in the Senate, not because of his position on flood control, but because of his opposition to the Ku Klux Klan. At the time, the Klan had friends in high places in Mississippi (including the governor's mansion). Although Percy was a planter who relied on cheap, abundant labor, he was no friend of the Klan and he denounced its terrorist tactics against African Americans.

Years later, Percy would gain a different kind of recognition. In 1934, despite Percy's loss of his Senate seat, the Mississippi legislature christened its first state park Leroy Percy State Park. Located near Hollandale, Mississippi, just south of Greenville and not too far from the banks of the Mississippi River, the park featured alligators, artesian springs, and ancient live oaks dripping with Spanish moss. As it became a popular place, thousands more acres of land were added to the Mississippi state park system. Percy, who relished the outdoors and who had accompanied President Theodore Roosevelt on his famous bear-hunting trip—the trip that led to the invention of "Teddy Bears" —would have been pleased. The venue for the hunting trip would eventually become the Holt Collier National Wildlife Refuge, named for Roosevelt's African American hunting guide.[65]

Soon after the creation of Leroy Percy State Park, one of Sir Willcocks' recommendations took root. Congress probably did not have Willcocks in mind, though, when it created the Mark Twain National Forest in 1939, in an area of the St. Francis River basin that had been intensively logged and then severely flooded in 1913. There, urban development was precluded, eventually allowing the forest to trap and filter floodwaters, and then release them slowly back into the river—just the type of natural flood control reservoir that Willcocks had advocated.

Meanwhile—Percy's legacy and future national forests aside—implementation of the Flood Control Act of 1917 was well underway by the 1920s. Federal navigation and flood control levees—massive walls of earth up to eighteen feet wide—stretched more than sixteen hundred miles along the lower river. As immense as they seemed, the levees would be no match for the great flood of 1927.

3

THE FLOOD OF 1927

SHELTERED BY IMMUNITY, THE CORPS VENTURES BEYOND THE "COLOSSAL BLUNDER" OF THE LEVEES-ONLY POLICY

The roaring twenties! It was a decade of celebration. After the floods of the previous decades had subsided and the armistice of 1918 ended hostilities on the Western Front, the nation was ready to put its losses behind and move forward. New technologies and a skyrocketing stock market made the future look bright. Superstars like baseball great Babe Ruth, with his astonishing sixty home-run season record, and trans-Atlantic pilot Charles Lindbergh made people dream of greatness. "Talking pictures" were all the rage in movie theaters throughout the country, and the first radio networks—NBC and CBS—were created. Although the Eighteenth Amendment (ratified in 1919) to the U.S. Constitution prohibited the manufacture and sale of alcohol, it didn't dampen the enthusiasm. The party just moved into the speakeasies, where flappers danced the "Black Bottom" and the "Charleston," and jazz aficionados listened to Duke Ellington, Louis Armstrong, and Jelly Roll Morton. Meanwhile, suffragists began to exercise the newly minted right to vote, ratified by the U.S. Congress in 1920.

The 1920 census was a benchmark as well. For the first time in history, the number of Americans living in cities surpassed those in rural areas. Cities offered more jobs and access to more products and social opportunities. Advertisers called for a Model T in every driveway, and Henry Ford's standardization of mass production processes at his Highland Park plant, where a new car came off the assembly line every sixty seconds, made automobiles more widely available and affordable.

But the era's exuberance did not quite reach rural America, especially in the South. In contrast to the prevailing trend of urbanization in the North, 70 percent of Southerners still lived in rural areas, and over 40 percent of all workers in the South were farm laborers. Farmers throughout the nation suffered after World War I, as agricultural markets shrank and prices collapsed, but Southerners were especially hard hit. By 1929, the boll weevil—a pest that appeared in Georgia in 1916 and rapidly spread throughout the South—had become, in effect, the largest consumer of raw cotton. Louisiana's agricultural sector faced an additional threat. Its sugar industry was nearly decimated by mosaic disease, a virus that stops plant growth by attacking chlorophyll in the sugarcane. Per capita income for farm workers fell to less than half of non-farm income.[1] While their urban neighbors were enjoying listening to dance music, comedians, and baseball games on the radio, 90 percent of American farms lacked electricity. As farm incomes plummeted and indebtedness rose, the proportion of farms with access to a telephone decreased over the course of the decade.[2]

Southerners were perceived as backwater bumpkins by Northern city-dwellers. Journalist H. L. Mencken, known by his contemporaries as the "Sage of Baltimore," skewered the South with his pen. He labeled the region the "bunghole of the United States"[3]—a place "almost as sterile, artistically, intellectually, culturally, as the Sahara Desert." It was not long before Mencken was unable to set foot in Dixie without an armed escort (assuming he ever wanted to set foot in Dixie, that is).[4] Other writers criticized the region's illiteracy, racism, disease, peonage, and political corruption in less colorful, but no less damning, terms.[5]

Of course, none of these attributes was limited to the South. Political corruption, in particular, was rampant during the 1920s, even at the national level. During the previous decade, President Woodrow Wilson had accomplished many notable things, especially during his second term in office, which saw the conclusion of World War I and the passage of the Flood Control Act of 1917. Republican President Warren G. Harding, elected in 1920, promised a "return to normalcy," which in his mind meant rolling back the perceived excesses of governmental activism. But Harding would go down in history as having condoned other kinds of excess. Known for his conservative, laissez-faire attitude, Harding took a "hands off" approach to governing not only the country but also his own cabinet, and his cronies took advantage of the president's managerial style. Harding's administration was riddled with scandals, the most notorious of which was the Teapot Dome Scandal.

Oil fields on federal public lands in California and at the Teapot Dome in Wyoming—named for a teapot-shaped rock situated above the oil deposit—had been reserved by Presidents Wilson and Taft as emergency naval supplies. But Albert B. Fall, Harding's Secretary of the Interior, convinced Edwin Denby, the Secretary of the Navy, to give control of the oil reserves to Interior. An executive order issued by President Harding paved the way for the transfer. Fall turned around and leased the oil production rights at Teapot Dome to Harry Sinclair's Mammoth Oil Company. In return for the Teapot Dome and the California leases, Fall received $400,000 in gifts from Sinclair and other oilmen. Fall's much-improved standard of living eventually gave him away, and investigations, indictments, and convictions soon followed. Fall paid $100,000 in fines and served a year in jail for bribery, giving him the distinction of being the first cabinet member to be imprisoned for misconduct in office. Sinclair was imprisoned for contempt of Congress and jury tampering. As for the Teapot Dome lease itself, the U.S. Supreme Court ruled that it had been unlawfully obtained and therefore was invalid.[6] For decades—indeed, until the Watergate scandal of the 1970s—Teapot Dome was widely known as the "greatest and most sensational scandal in the history of American politics."[7]

* * *

AMONG A DECADE of remarkable years, 1927, in particular, stood out. Historian Allen Churchill memorialized it in his book *The Year the World Went Mad*. According to Churchill, 1927 was the pinnacle of "the Era of Wonderful Nonsense." Events ranged from the sublime to the disastrous. Charles Lindbergh enjoyed international acclaim when he became the first pilot to complete a non-stop solo flight across the Atlantic Ocean in his single-engine plane, *The Spirit of St. Louis*. Secretary of Commerce Herbert Hoover experimented with the newly invented television, which, with its tiny two-by-two-and-a-half inch screen, generated a great deal of excitement, although the *New York Times* believed that its "Commercial Use [was] in Doubt." The same year, Italian immigrants and self-proclaimed anarchists Nicola Sacco and Bartolomeo Vanzetti were put on trial and condemned to the electric chair for the murder of two men in Massachusetts in a highly publicized verdict that many people, especially in Europe, thought was a product of prejudice and procedural irregularities.[8] The biggest event of the year, however, was the Great Mississippi River Flood. The flood hit the American South especially hard, but it had nationwide repercussions.

Faith in Inexhaustible Resources

LIKE THE OIL under the Teapot Dome, natural resources were being developed at a break-neck pace during the 1920s. While income from annual agricultural crops like cotton, corn, and sugarcane dropped, the timber industry thrived, especially in the South. Post-war construction boomed, demand for timber rose, and by 1921, an area the size of Georgia, Alabama, and Mississippi combined had been completely denuded of trees. Much of the logging took place in floodplains and coastal areas, leaving bare ground to erode into the rivers and bays. After the virgin stands of timber were gone, lumber mills began processing waste byproducts into boxes, pulp, and paper. Although some areas were replanted with fast-growing new stands of pine seedlings, many were not.[9]

Water resources and waterways were exploited as well. In the early 1900s, engineers and government bureaucrats assumed that the same structures that would improve navigation and produce water supplies and hydropower would also contain floods and provide drainage.[10] The Corps had long advocated more and bigger levees for navigational purposes, and other agencies climbed on board, hoping that levees would be an effective means of achieving multiple objectives—thereby providing something for everyone. Federal agencies were highly fragmented, and jealousies raged between the Departments of Interior and Agriculture, and between Interior and the Corps of Engineers. A new federal agency, the Federal Power Commission, established in 1920 with the passage of the Federal Power Act, complicated matters even further by assuming responsibility for hydropower. According to historian Donald Pisani, "modern water politics was born in the 1920s," complete with logrolling, pork-barrel funding, and backroom deal-making.[11]

Thanks to the floods of 1912 and 1913, and the Flood Control Act of 1917, the Corps had begun to gain the upper hand in terms of congressional appropriations, authority, and territorial control in the Mississippi River basin. Until 1917, Congress insisted that federal money could be used for levees and other measures *only* insofar as they were related to navigation, *not* to protect land from flooding.[12] Congress relented somewhat in the Flood Control Act of 1917, the first federal law that explicitly appropriated money for river infrastructure developed for non-navigational purposes.[13] It allocated $45 million for flood control work between the mouth of the Ohio and the mouth of the Mississippi. The project was supervised by the Mississippi River Commission, an entity formed primarily of officials from the Corps of Engineers.

Despite the 1917 act, levees remained a local affair, for the most part. Local entities were required to secure the necessary rights-of-way for levees, to contribute to the cost of levee construction, and to maintain the levees after they were built.[14] The prevailing sentiment at the time favored local control over land-use planning and flood control. Critics of federal power continued to question whether the central government had the authority—much less the responsibility—to shield private property from flooding.[15]

The state legislatures of Louisiana and Mississippi created specialized units of local government to coordinate flood control activities across parish or county lines. These units were authorized to establish levee districts, to appoint officials to inspect levees and drainage ditches, and to compel landowners within the floodplain to conduct some of the necessary construction and maintenance work. The inspectors were also authorized to impose fines upon landowners who neglected their flood control duties. Perhaps most remarkably, inspectors could even fine landowners for failing to conscript their employees into service to build the levees.[16]

In 1926, the Chief of the Army Corps of Engineers looked out over the expanse of federal, state, and local levees along the Mississippi and declared that the fortifications were "now in condition to prevent the destructive effects of floods."[17] His confidence was misplaced.

April is the Cruelest Month

IN THE WINTER and spring of 1927, rain fell in sheets in the lower Mississippi basin. Five separate storms rolled through, each one more powerful than anything residents had ever experienced before. Instead of the promise of resurrection, Good Friday brought more rain. Storms pummeled a one-hundred-thousand-square-mile area from the confluence of the Mississippi and Ohio rivers at Cairo, Illinois, to the Gulf of Mexico. New Orleans broke all records with an astonishing fifteen inches of rain in only eighteen hours.[18]

Two men standing on porch
of house surrounded by water,
Greenville, Mississippi, April
30, 1927
*Source: Mississippi Department
of Archives & History*

The Mississippi River Commission assured the residents that all
levees built to government specifications would hold. The commission was wrong. Just before Easter Sunday, on April 16, 1927, the first
of seventeen major crevasses occurred in a federal levee near Dorena,
Missouri, thirty miles below Cairo, Illinois.[19] The raging flood waters
gouged out twelve hundred feet of the levee. Author John Barry gives a
vivid description of the event:

> The river poured through the breach, tearing down trees, sweeping
> away buildings, and destroying faith. . . . The Mississippi was three
> miles wide between the levees, darker and thicker and more wild than
> any man, red, black, or white, had ever seen it. Detritus of the flood
> —tree branches and whole trees, part of a floor, a roof, the remains of a
> chicken coop, fence posts, upturned boats, bodies of mules and cows
> —raced past.[20]

A few days later, a crevasse developed at the Mounds Landing levee,
around three hundred miles downstream from Dorena and only twenty
miles north of Greenville, Mississippi.[21] The levee, which had been
constructed by a local levee district in 1867, was situated in a particularly
vulnerable spot on a ninety-degree river bend, and Greenville residents
feared that it would be no match for this flood. Black sharecroppers
and workers from nearby plantations were forced at gunpoint to stack

sandbags beside and on top of the levee in an attempt to keep the river from pouring over. To receive food rations, black men were required to wear tags identifying themselves as laborers and showing which plantation they "belonged to." National guardsmen patrolled the levee camp to prevent them from leaving and to keep them working.[22] When a nineteen-year-old boy asked for a rest break, he was pistol-whipped by the police. When Mrs. Nancy Clark Peters objected to her husband's forced conscription, she was arrested and thrown in jail. When the levee failed, the current swept away and drowned countless men—mostly black.[23] The need for extra hands during the flood was only part of the equation. The *Chicago Defender*—the first black newspaper to have a circulation over one hundred thousand—ran exposés about the brutal tactics of Southern planters who feared that Northern employment agents would lure African Americans away from the region, leaving the South without cheap labor.[24]

It soon became apparent that more hands were needed to combat the flood. One of the chief agitators for federal flood control in the wake of the 1912–1913 floods, former U.S. Senator LeRoy Percy—who owned twenty thousand acres near Greenville—persuaded the governor to send convicts from the state penitentiaries to assist. But the Mississippi River swept the sandbags away almost as quickly as they were laid, along with a mile-wide section of the levee itself and many of the workers. The *Memphis Commercial-Appeal* reported: "Thousands of workers were frantically piling sandbags . . . when the levee caved. It was impossible to recover the bodies swept onward by the current at an enormous rate of speed."[25]

The break widened until a one-hundred-foot-high wall of water cascaded over the delta. Within just ten days, the deluge submerged one million acres beneath ten feet of water. And still, the river continued to gush relentlessly through the Mounds Landing gap for months. In all, the levees ruptured in 120 places. The flood covered nearly eighteen million acres in seven states—an area nearly twice the size as the state of Maryland. At its widest point, just north of Vicksburg, Mississippi,

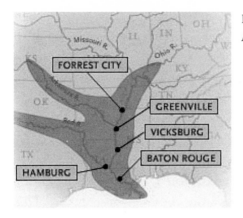

Flood of 1927
Map by Public Broadcasting Service

the swollen river formed an inland sea nearly one hundred miles wide.[26] The flood brought particular hardship to black refugees. In many cases, government workers and volunteers denied them shelter, food, and medicine. *The Workers Vanguard* described how the displaced blacks were treated in the racially segregated Red Cross camps: "Black families lived in floorless tents in the mud without cots, chairs or utensils, eating inferior rationed food. . . . Typhoid, measles, mumps, malaria and venereal diseases ran rampant among destitute tenant farmers and mill workers already weakened from illnesses endemic to poverty, such as tuberculosis and pellagra."[27] According to author John Barry, "the stench was unbearable."

True to his conservative leanings, President Calvin Coolidge (1923–1929) was reluctant to involve the federal government. Early rescue and relief efforts were handled by private individuals and organizations, especially the American Red Cross, which had been founded in 1881 by Clara Barton, a well-known humanitarian and suffragist. But the state governors demanded more assistance. Movie theaters and radio stations added pressure by sensationalizing the worst of the horror stories coming from the flood-ravaged states. Once again, ever ready with his pen and printing press, H. L. Mencken dramatized public sentiment in typical acerbic fashion: "Nero fiddled, but Coolidge only snored."[28]

Coolidge eventually conceded some ground by appointing Secretary of Commerce Herbert Hoover to manage the disaster response. The U.S. military provided seaplanes from Pensacola Naval Air Station for daily reconnaissance missions to inspect levees, locate refugees, map out rescue routes for watercraft, and provide food and medical supplies. Services were limited, however, and in many cases, black workers and refugees were denied evacuation assistance. They were also cut off from supplies by plantation owners and overseers who wanted to hoard provisions for their own families and friends.[29]

Estimates vary, but government sources reported that the flood caused between $230 and $360 million in property damage (up to $4.5 billion in 2010 dollars).[30] The hardest hit states were Arkansas, Mississippi, and Louisiana. Arkansas claimed the most extensive property damage, with over two million acres of agricultural lands and nearly sixty thousand houses inundated. In Little Rock, a twenty-one car train laden with coal toppled into the Arkansas River, along with the bridge it had been resting on.[31] Arkansas was not alone in the loss of the Little Rock Bridge; the flood destroyed every bridge in its path for a thousand miles.[32]

Mississippi experienced the highest death toll.[33] Officially, the federal government reported that five hundred people died in total, but a disaster expert who visited the area estimated that over one thousand perished in Mississippi alone.[34] Hundreds of thousands of flood-displaced survivors took up residence in box-cars, make-shift tents, and Red Cross camps; many stayed for over six months.[35]

New Orleans was spared the worst of the flood damage. Levee breaks upstream in Mississippi and elsewhere had lessened the threat to the city by dispersing the water across the floodplain before it reached the delta. Even so, as a precaution, New Orleans bankers, businessmen, and politicians convinced President Coolidge and the Corps to blow up the Caernarvon levee to force the water out through the manmade crevasse and away from the city. That particular levee protected a poor, rural area known as St. Bernard Parish (which would be ravaged again, almost a century later, by floodwaters associated with Hurricane Katrina). The National Guard carried out the evacuation order and forced ten thousand parish residents to leave before the Corps blasted the levee and their homes were inundated. Refugees were herded into a warehouse in downtown New Orleans and segregated into two camps—whites on the fifth floor and blacks on the sixth. Eastern St. Bernard Parish and part of Plaquemines Parish were completely destroyed.

Afterward, a reparations commission was established to administer the residents' claims for property damage. Most received only pennies on the dollar for their losses.[36] In one case, Claude Foret, the general manager of the St. Bernard Motor Company, sued the Orleans Board of Levee Commissioners to recover his $2,625 salary that was lost during the inundation of the parish. As a result of the flooding, Mr. Foret was out of work for eight months. His claim was met with stony indifference, first by the commission, and then by the Louisiana Supreme Court. According to the court, in the event of public emergency, any damages occasioned by the government's destruction of a levee were *damnum absque injuria*—loss that falls short of legal injury. Additionally, the court found that flood reparations were intended to remedy property damages, *not* consequential damages such as the loss of salary.[37]

The Louisiana Supreme Court was equally unimpressed with claims brought by the St. Bernard fishermen. The Alfred Oliver Company, which, as the court noted, owned and operated "a large fishing boat fully equipped with seine and trawl," lost $160 per week in net profits over a total of eighteen weeks due to the inability to fish while excess floodwaters gushed through the dynamited levee. In an opinion issued on the same day that Mr. Foret lost his case, the court found that the fishermen had no proprietary interest in the fish located in public waters owned by the state and therefore were entitled to no relief whatsoever.[38]

Although his clients walked away empty handed, Leander H. Perez, the attorney who had represented both the Alfred Oliver Company and Mr. Foret, continued to enjoy and exploit his reputation as one of the most powerful, and most corrupt, political forces in Louisiana.

Politicians backed by Perez were virtually certain to win, due to Perez's unprecedented ability to falsify voting records, sometimes using names of celebrities such as Babe Ruth and Charlie Chaplin to pad the voter rolls. In 1928, the year after the great flood, Perez threw his support behind Huey Pierce Long, Jr., in Long's successful campaign for governor. Perez later defended Long when the governor faced impeachment charges before the state legislature for corruption, bribery, and even blasphemy, which was deemed an offense of a political character and therefore grounds for impeachment. With Perez's assistance, Long prevailed against the impeachment charges and remained in power for several years, both as governor and later as a U.S. senator (he succeeded Senator Joseph Ransdell, principal architect of the 1917 Ransdell-Humphreys Flood Control Act). Long's unique abilities in creative means of retaliating against political opponents got him killed in 1935, when Dr. Carl Weiss, the son-in-law of one of Long's long-time adversaries, shot Long at close range at the state capitol. Leander Perez, on the other hand, continued to influence Louisiana politics well into the 1960s, so much so that a major thoroughfare in St. Bernard Parish was named after him. In 1999, however, parish officials renamed the street in an effort to distance themselves from Perez's legacy of corruption and racism.[39]

A Federal Response to the Main Street Myth

IN SPITE OF the limited media coverage available at the time—there were no twenty-four-hour news networks, and, indeed, there was no commercial television at all—the 1927 flood left an indelible mark on the nation's political and social landscape. The flood motivated the public to re-examine long-standing perceptions of the limited responsibilities of the federal government. The myth that powerful individuals and local politicians could effectively manage floods was debunked. No longer could the government stand by and let "Main Street save Main Street," as then-Secretary Hoover explained.[40] Bankers, realtors, and everyday citizens cried out for federal leadership, technology, and financial resources to prevent floods and to remediate their devastating effects. The U.S. Chamber of Commerce warned Congress that the federal government must undertake the necessary work, lest the country return to a "great waste" from Cairo to the gulf.[41]

Flood control was the most pressing issue before the Seventieth Congress, which sat from 1927 to 1929.[42] Congressional members quickly recognized that the problems were two-fold. First, there was an absence of federal leadership. As Congressman Edward Denison of Illinois pointedly observed, "[t]he Federal Government has allowed the people . . . to follow their own course and build their own levees as they choose and where they choose until the action of the people of one State has thrown the waters back upon the people of another State, and vice versa."[43] Moreover, the little guidance that Congress had provided to date was not helpful. As Congressman Robert Crosser of Ohio noted, the "levees-only" policy was not the right sort of federal leadership:[44]

> Many millions of dollars have been spent in an effort to control floods in the Mississippi Valley. Most of the work has been worse than wasted, for it has done much harm instead of good. . . . We have spent many millions of dollars to build levees; that is, great embankments alongside of and a little distance from the natural banks of the river, and the result has been that every flood has been more disastrous than the floods which preceded it.[45]

Like Congressmen Denison and Crosser, Gifford Pinchot, who had served as the chief of the forest service under President Theodore Roosevelt, criticized the Coolidge administration and particularly the Corps of Engineers. He had spoken out against the Corps' "levees-only" policy long before 1927, and after the great flood he called it "the most colossal blunder in engineering history."[46] On surveying the flood damage, Pinchot concluded, "This isn't a natural disaster. It's

Mother Goose and flood relief, April 14, 1928
(editorial cartoon by J. N. "Ding" Darling)
Source: Courtesy of the Jay N. "Ding" Darling Wildlife Society

a manmade disaster."[47] Decades later, even the Army Corps of Engineers acknowledged that "while levees were indispensable, to depend on them for flood protection was suicidal."[48]

Despite the widespread zeal for reform, President Coolidge balked. He insisted the federal government should not be in the business of protecting people from acts of God, such as floods. "The Government is not an insurer of its citizens against the hazard of the elements," Coolidge remarked in his annual message to Congress.[49] He stubbornly maintained that local citizens must be charged with responsibility for the cost of flood control to ensure that they had a direct "pecuniary interest in preventing waste and extravagance."[50]

Proponents of a federal flood control package partially overcame Coolidge's resistance by sweetening the pot with requirements for state and local funding contributions.[51] With the urging of his fellow Republicans, who feared that a veto would have serious political repercussions, Coolidge held his nose and signed the Flood Control Act of 1928 (sometimes known as the "Jadwin Act" because it adopted into law a Corps' flood protection plan of the same name).[52] Coolidge had badly underestimated the voters' sentiments, however, and the damage to his public image from his cavalier handling of the 1927 flood and other missteps could not be undone. Once the political logjam over federal flood control broke, Coolidge was swept out of office. Another Republican, Herbert Hoover, described by author John Barry as the "logistical genius" who had orchestrated the rescue and rehabilitation of nearly a million destitute people in the Mississippi River valley during the 1927 flood, rode the wave all the way to the Oval Office in 1929.[53]

According to Louisiana Senator Joseph Ransdell, the 1928 act was the most remarkable piece of water-related legislation "since the world began."[54] Representative Frank Reid of Illinois announced that it was "the greatest piece of legislation ever enacted by Congress," bar none.[55] Former Senator LeRoy Percy must have been quite satisfied that his many years of both overt and behind-the-scenes lobbying had paid off, and that his beloved state of Mississippi would soon see federal dollars, instead of more water, flowing toward it.

The Flood Control Act of 1928 declared that the federal government, through the Corps of Engineers, would take responsibility for flood control in the Mississippi basin.[56] The legislation incorporated the positions embraced by both Congressman Denison (calling for federal leadership) and Congressman Crosser (calling for more infrastructure). Still, the Corps' new duties were primarily structural, requiring the construction of more federal levees. Importantly, in addition to levees, the law also called on the Corps to construct "spillways," "floodways," and

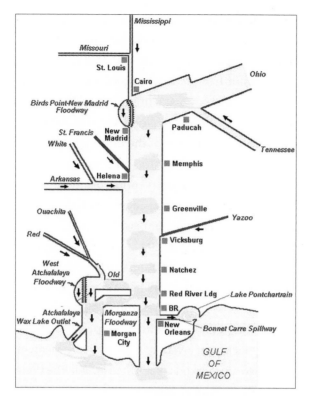

The Mississippi River and Tributaries Project floodways
Source: Mississippi River Commission

reservoirs to replicate the flood-taming function of natural floodplains. Moreover, the act authorized federal flood control activities beyond the Mississippi basin—in the Sacramento River basin in California—in large part to get support from Westerners in Congress. Armed with its new authority, the Corps got busy building dams, spillways, floodways, and other flood control structures well beyond the traditional "levees-only" strategy. The Corps designed its plan, known as "the Mississippi River and Tributaries Project" (MR&T Project), to withstand what it called the "project flood"—the levels of flooding experienced in 1927, plus a margin of safety.

Yet even as Congress expanded the federal government's flood control authority, Congress was careful to limit the federal government's responsibility in case flooding occurred despite (or even because of) the Corps' efforts. In particular, it was not willing to pay for any damages that might result from federal flood control activities. The act, therefore, immunized the government from liability "of any kind . . . for any damage from or by floods or flood waters at any place."[57] This provision is an explicit statement of the notion, long embedded in the law, that "the King can do no wrong," and even if he does, he cannot be sued. In any event, judicial intervention has only rarely curbed the Corps' activities; deference to the agency runs high, especially when it comes to complex river management issues.[58]

By today's standards, the 1928 act was a modest measure, but in fiscal terms it was more expensive than almost anything the federal government had ever undertaken except World War I. The levees, reservoirs, and outlets authorized by the act cost around $325 million.[59] Even more importantly, by setting a precedent for widespread federal involvement in what had long been perceived as the primarily local affair of flood prevention, the 1928 act marked a paradigm shift in the division of labor among federal, state, and local governments.

* * *

WHILE THE GREAT Mississippi River flood of 1927 set the stage for greater federal involvement and stimulated the addition of floodways and other engineered flood control devices to the levee system, the government made no attempt to prevent people from moving into floodplains, whether natural or one of the newly authorized federal floodways. In fact, as the next chapter explains, misguided federal policies actually discouraged people from leaving riverfront lands, even after the areas had been designated as official floodways for the storage of excess flows. What would happen during the next great flood, when levees alone weren't enough, and the new floodways were pressed into service?

4

THE FLOOD OF 1937

THE CORPS BUILDS FLOODWAYS

After Congress passed the Flood Control Act of 1928, the nation embarked down a new path. In adopting a flood control strategy for the lower Mississippi, Congress had drawn upon the expertise of the nation's best and brightest engineers, who had produced more than three hundred alternative plans. Congress settled on the "Jadwin Plan," authored by the Corps' chief engineer, Major General Edgar Jadwin. A compromise, the plan attracted its share of opponents, including those who were in the path of the newly designed floodways and spillways. When the Great Depression hit in 1929, just one year after enactment, other critics worried that the federal government simply could not afford such expensive efforts. But President Hoover convinced Congress to continue funding Mississippi River flood control projects despite the budgetary crisis by characterizing them as relief programs for the unemployed. Hoover was otherwise opposed to unemployment relief programs, and he refused to authorize direct federal aid for the unemployed. The voters wanted more aid, and, like Coolidge before him, Hoover was voted out of office in 1933 after only one term.

The Depression was the result of a "perfect storm" of events. In the agricultural sector, new technologies and government incentives during and after World War I stimulated a tremendous increase in grain production. Before long, there was a glut in the global wheat market. Prices plummeted. A bushel of wheat brought only seventy-five cents in 1929 and twenty-five cents in 1930, down from $2.25 just a few years earlier. The ability to bring in a bumper crop did the farmers little good when prices were so low.[1] The commercial sector soon followed suit. Bankers, businessmen, and city dwellers felt the pain on Black Tuesday, October 29, 1929, when the stock market crashed. Thousands of businesses and banks went under, and up to fifteen million Americans— one-quarter of the workforce—were out of work. The nation descended into the worst economic collapse in U.S. history.[2]

Weather played a role in prolonging the Depression, but this time it was the *lack* of precipitation that wreaked havoc. Shortly after the 1927 flood devastated the South, the 1930s ushered in the most severe drought that the country had ever experienced. Journalist and author Timothy Egan described it as "the nation's worst prolonged environmental disaster," with the land itself becoming "an active, malevolent force," as black dusters, formed of loose, dry topsoil picked up by incessant winds, whirled across the land.[3]

Congress took note in the spring of 1935, when tons of dust from the southern plains swept through the Midwest to Washington, D.C. The dust eclipsed the sun and turned the air a dark copper color, just as a Senate committee was considering a proposal to create a national Soil Conservation Service. Congress promptly declared soil erosion "a national menace" and authorized the formation of the SCS.[4] The SCS and related programs attempted to rehabilitate the Dust Bowl by changing the basic farming methods of the region. Conservation measures such as seeding areas with grass, planting shelter belts of trees to break the wind, rotating crops, and using contour plowing would not only prevent erosion but would also help store excess water and prevent it from flooding the land.[5] Soil conservation was only one piece of President Franklin D. Roosevelt's grand "New Deal" plan. President Roosevelt, who had ousted Hoover in the 1932 election, made public welfare a matter of federal concern, from the farm fields to the coal mines to the soup lines.[6] The Roosevelt administration put people to work in soil conservation districts, parks, and forests, and on sewage treatment plants, dams, and other water-related projects.

Even as New Deal initiatives exploited the broadest possible array of economic outputs from river basins, from hydropower to irrigation, federal water policy became increasingly disjointed. Bureaucratic rivalries deepened, as cabinet officials such as Interior Secretary Harold Ickes and Agriculture Secretary Henry Wallace competed for public works spending. Meanwhile, Congress expanded the power of the Corps over both navigation and flood control, and created the Tennessee Valley Authority in 1933 to oversee water resources development in the Southeast. Basin-wide, coordinated planning became nearly impossible — there were too many fingers in the water resources pie. New Deal policies spawned massive multi-purpose federal projects by treating them as local job relief rather than integrated parts of a national whole, and water-related programs became ever more fragmented.[7]

During the New Deal, Congress added another layer to the complicated assortment of authorities by passing the Flood Control Act of 1936. This act is particularly notable because for the first time Congress explicitly recognized federal responsibility for flood control *nationwide*, not just in the Mississippi basin. In it, Congress proclaimed that, like soil erosion, "destructive floods . . . upsetting orderly processes and causing loss of life and property, . . . constitute a menace to national welfare."[8] To control this "menace," the 1936 act delegated broad discretion to the Corps to construct any flood control project it chose, so long as Congress agreed to appropriate the necessary funds. It allowed the Corps to proceed whenever "the benefits to whomsoever they may accrue are in excess of the estimated costs."[9] Beginning in 1936, the Corps would spend billions of dollars on dams, reservoirs, spillways, levees, dikes, and other structures for flood control and related purposes.[10]

* * *

WHEN CONGRESS ADDED floodways to the nation's arsenal of levees and other measures to combat flooding, it knew that farms and rural communities had already been established in areas that the Corps would likely select as emergency overflow areas. Invoking the federal governmental power known as *eminent domain*, Congress ordered the Corps to condemn (and pay for) the necessary property rights. In essence, eminent domain is a forced sale from an unwilling seller to the government. It allows federal, state, and local governments to assemble large tracts of land not otherwise for sale or available. The newly acquired lands can serve a variety of "public uses," such as road building, railroad development, power line installation, and other public projects. They can also be pressed into service for the location of dams, floodways, and spillways. The power of eminent domain has been recognized for hundreds of years in the United States, England, and other nations, sometimes without even requiring the government to pay for the privilege. However, the drafters of the U.S. Constitution's Fifth Amendment explicitly required the government to pay "just compensation" (generally, fair market value) when it takes over private property through the exercise of eminent domain. Although seemingly straightforward, the condemnation process can be contentious, difficult, and time consuming.

After the flood of 1927, Congress ordered the Corps to exercise eminent domain in at least two situations. First, Congress recognized that in some cases where the Corps planned to construct levees on only one side of the river (due to cost or engineering difficulties), the opposite side of the river might be exposed to an increased risk of overflow and damage. In those cases, Congress ordered the secretary of war and the chief of engineers to condemn the vulnerable lands through eminent domain. Second, Congress ordered the federal government to condemn property designated for use as spillways and floodways. In both cases, the law authorized the Corps to purchase the lands outright, and the affected landowners would presumably use the funds to move out of harm's way. But the law also allowed the Corps to purchase something less than full title (at a lower cost) known as *flowage easements*. Easements are a type of shared property interest in which the landowner retains title to the property, but allows the easement holder to use the property for specified purposes (as where, for example, one neighbor sells another the right to drive through his or her property as a shortcut to a main road). In the case of flowage easements, the Corps acquired in advance the legal right to "use" otherwise private property from time to time for the storage of floodwaters. It doesn't take a legal expert to see that this type of sharing may work well in theory, but will be disastrous in practice.

Why would Congress continue to support floodplain settlement by providing only a partial buy-out of some endangered properties? And why would any landowner remain on lands clearly designated as "floodway" on the map? Perhaps the Great Depression played a role, simultaneously limiting the federal purse and clouding the judgment of people eager to take any money the government offered, particularly if they could still remain on the land. Perhaps some landowners dedicated low-lying areas to farming, and built their homes on safer, higher ground. Or perhaps people simply could not believe that a disaster on the scale of the 1927 flood would ever happen again. After all, when the Corps began to purchase flowage easements, the nation was plagued by extreme drought and devastating dust storms, not by flood.

The Walled City of Cairo

THE FLOOD CONTROL legislation of 1928 and 1936 must have seemed like manna from heaven to Cairo, Illinois. Situated on a narrow peninsula where the Ohio and Mississippi rivers join, Cairo is frequently at the mercy of floodwaters.

The town's founders believed that the risks of settling in a floodplain were outweighed by the commercial advantages provided by the water-borne trade supported by the two great rivers. And they believed they could wall off the city with levees and floodwalls sufficient to protect it.

In 1851, a visiting Lieutenant-Colonel Arthur Cunynghame opined that Cairo, Illinois, despite periodic flooding by the Mississippi and Ohio rivers, would one day compare favorably to the city of the same name in Egypt: "Cairo, [Illinois,] though now insignificant, may in a few years excel, both in wealth and in size, as it speedily will in intelligence, its older namesake, Cairo on the Nile, whose propensities to overflow her banks are the same as the Mississippi." Likewise, in 1859, a commentator from France, Jules Rouby, marveled at the "American ability" that would likely "triumph over [the rivers] by means of perseverance, labor, and expenditure of money." Rouby was struck by the city's aspiration to use its "incomparable geographic situation" to become a major commercial crossroads.[11]

Just below the confluence of the Ohio and Mississippi rivers, a choke point threatened to back up the rivers when in flood, thereby overwhelming upstream Cairo from multiple directions. On the Illinois side, steep bluffs wall in the river from the east. To the west, on the Missouri side, riverfront levees up to sixty-feet high (also called *frontline levees*) lined the Mississippi for fifty-six miles from Birds Point, Missouri, down to New Madrid, Missouri. As part of the Mississippi River and Tributaries Project authorized after the 1927 flood, the Corps set to work downstream from Cairo. Its goal was to relieve the pressure of floodwaters against the old frontline levees and to lower the flood crest of the Mississippi River at Cairo by up to seven feet. First, the Corps built a thirty-six mile stretch of *setback levees* that ran about five miles to the west of, and parallel to, the frontline levees in Missouri. Next, to relieve the pressure at extremely high river stages, the Corps designed a series of *fuse plugs*—sections of the riverfront levees that would be about two feet lower than the rest of the levee. When necessary, the Corps would direct the swollen river through the fuse plugs onto about 130,000 acres of rural lands situated between the levee systems, an area named the "Birds Point-New Madrid Floodway." Finally, the Corps began to condemn property and create flowage easements in the newly designated floodway.

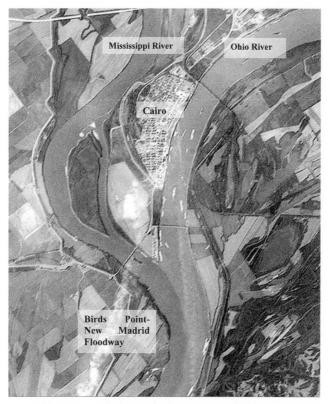

Confluence of Ohio and Mississippi rivers at Cairo, Illinois
Source: NASA, Earth Observatory

The scheme would protect the citizens of one state (Illinois), but at the expense of the citizens of another (Missouri). It elevated urban over rural interests. At that time, in the words of the Mississippi River Commission, Cairo, Illinois, was "a lavish and bustling river town with a population exceeding 15,000."[12] According to plan, Cairo would be spared from severe floods, while the risk would be shifted to rural landowners and farmers in Birds Point, Missouri. Congressman Dewey Short explained his opposition to the Jadwin Plan, stating that he and his Missouri district did not "want to see southeast Missouri made the dumping ground to protect Cairo, much as we love Cairo."[13] Located in a sacrifice zone, these rural landowners and farmers knew that they could be flooded at any time. Some would be bought out completely by the federal government. Others would remain behind, subject to flowage easements held by the federal government.

The process of condemning property in the floodway proved to be complicated and time consuming. In fact, the Corps would not complete its condemnation work until 1942. By that time, it had purchased flowage easements for more than 106,000 acres, at an average cost of $17 per acre.[14] In the meantime, the Ohio River took charge. Not content to wait until the Corps was ready for it, the river threatened to flood in 1937, just ten years after the devastating Mississippi River flood of 1927. By this time, the Corps had substantially completed both riverfront and setback levees. But the Corps had not yet created the fuse plugs. The Corps held off on that essential piece of work because it was bogged down in settlement discussions and litigation over its condemnation of flowage easements.

Without the fuse plugs, there was no way to relieve the pressure at Cairo and to direct the raging river into its new floodway. No way, that is, except by the use of dynamite.

The situation was dire in January 1937, as the Ohio surged to a flood stage higher than any previously recorded. As water levels threatened to overtop the Cairo levees, the Army Corps of Engineers acted decisively: it blasted an emergency escape hole in the riverfront levee just downstream of Cairo, allowing the Mississippi to gush out of its channel and into the Birds Point-New Madrid Floodway. The Corps acted just in time, saving Cairo. According to government records, the floodwaters reached a height of 59.6 feet, stopping less than a half-inch below the top of the city's sixty-foot levee.[15]

But downstream, at least three thousand (and up to twelve thousand, by some estimates) tenant farmers and sharecroppers occupied the floodway. According to a *New York Times* article, the waters "poured in with such force that pilots in observation planes reported seeing houses, barns and other farm buildings crushed in the fertile valley."

One historian charged that the Corps deliberately flooded the poorest people and the richest cotton land in its effort to protect Cairo. According to some reports, the Corps gave the residents but three days' notice to move their families, possessions, and livestock out of the path of the torrent soon to be directed their way. On the other hand, Corps' supporters claimed that "at least 500 persons had remained in the basin despite warning by government engineers that they did so at their own risk."[16] Recalcitrant or not, the remaining residents paid a high price.

Overall, the Corps was satisfied with its efforts. It concluded that the floodway had been "of material aid" in reducing flood heights at and above Cairo during the 1937 flood.[17] But even so, doubts lingered. Afterward, the Chief of Engineers, Major General Edward Markham, testified before a congressional committee that "I am now of the opinion that no plan is satisfactory which is based upon deliberately turning floodwaters upon the homes and property of people, even though the right to do so may have been paid for in advance."[18]

Although Cairo was spared, the 1937 flood ravaged the entire Ohio River valley, including parts of Pennsylvania, Ohio, West Virginia, Indiana, Kentucky, and Illinois. Damage estimates varied widely, reaching up to $250 million—about $3.3 billion in today's dollars.[19] NOAA describes the flood as one that "surpassed all previous floods during

Piling sandbags along the levee in
Cairo during the height of the flood of
1937 (photograph by Russell Lee)
*Source: Library of Congress, Farm
Security Administration collection*

the 175 years of civilized occupancy of the lower Ohio floodplain."
The flood, in combination with the Great Depression, threatened to
plunge many middle-class families into poverty. The federal Resettle-
ment Administration set up refugee "tent" camps, as documented by
acclaimed New Deal photographer Russell Lee and others. To help
flood victims resettle elsewhere, the administration offered loans and a
variety of desperately needed social services.

Cairo's population peaked at about fifteen thousand sometime during the 1920s. By the end of the century—beaten down by repeated
flooding, economic woes, racial turbulence during the struggle toward
desegregation, and other challenges—the population declined to about
thirty-six hundred.[20] But the levees and floodwalls continue to sur-
round the city, representing a multimillion dollar investment standing
in boldfaced defiance of the Mississippi and Ohio rivers.

*   *   *

JUST DOWNSTREAM, IN the Birds Point Floodway, the lure of the
rich alluvial soils proved too strong to resist. Over time, after the 1937
flood, farmers came back to cultivate corn, wheat, and soybeans. Almost
one hundred families settled in the floodway, even though many were
ineligible for mortgages or insurance because of the known flood risk.
Life returned to easy normalcy for generations of farmers. Though the
area did flood periodically, most residents learned to take to the high
ground and ride it out. A decade into the twenty-first century, the area's
septuagenarians looked back on a lifetime of farming in the shadow of
the Mississippi and Ohio rivers, cultivating some of the richest soil in
the world. They knew of nothing else.

But the spring of 2011 brought the unthinkable—record floods on
both the Mississippi and Ohio rivers that approached and, in some
places, exceeded the levels of the 1937 flood. To save Cairo and other
population centers up- and down-stream, the Army Corps of Engineers once again dynamited open the Birds Point Floodway—for the
first time since 1937. Each second, 550,000 cubic feet of water blasted

through a half-mile wide gaping hole in the levee. A wall of water fifteen feet high obliterated about ninety homes. Overnight, a two-hundred-square-mile lake submerged 130,000 acres of newly sprouted corn and knee-high waves of wheat.[21] The following day, a group of farmers filed a class action lawsuit against the Army Corps of Engineers. They admitted that the Corps had purchased flowage easements across their property, but claimed that the depression-era easement condemnations (with some subsequently-negotiated modifications) were not broad enough to cover the circumstances generations later, and sought hundreds of millions of dollars in damages for breach of contract and for an unconstitutional *taking* of their property (a theory described in chapter 10 under which property owners seek compensation from the government if its actions or regulations adversely affect their property values).[22] The court roundly rejected their claims, finding that the flooding did not exceed the scope of the flowage easements. Further, the court determined that the government's flood control projects provided the landowners with more benefits than harms, which included only two floods separated by nearly seventy-five years. At most, the court determined, the plaintiffs were entitled to bring a trespass claim for consequential damages associated with one-time crop damage.

* * *

NOTHING HAD CHANGED, it seemed, since the great flood of 1927. The Army Corps of Engineers and the courts continued to turn a blind eye to the dangers of flood control. And landowners continued to believe they could have it all—fertile soil and floodless floodplains. The subsequent floods of 1937 and of 2011 raised as many questions as they answered. Should people live and farm in the shadow of the Mississippi and Ohio rivers? Should the federal government countenance, and even encourage, such flood-prone settlements? The richness of the farmland is exceedingly high. Unfortunately, so is the risk of setting down roots right in the path of the overflow from two great rivers.

5

MID-CENTURY FLOODS IN THE
MISSOURI RIVER BASIN

CONGRESS PROMISES SOMETHING
FOR ALMOST EVERYONE

The Missouri River—the Mississippi's longest tributary—is just as prone to flooding as the Mississippi itself. While the Corps was busy erecting levees, floodways, and other structures throughout the nation under the auspices of the 1928 and 1936 Flood Control Acts, the Missouri River rose up from its banks in 1937, 1942, and again in 1944. By then, the balance of power over commerce and water resources development had shifted 180 degrees from state and local governments to the federal government. As law professor Rena Steinzor explains, "The federal government created an array of national institutions to govern everything from farm policy to higher education, from inner-city housing to social welfare programs, and from interstate highway systems to environmental protection."[1] No longer did the Washington bureaucrats hesitate to address issues like farming and flood control.

Congress devoted its attention to the Missouri in the Flood Control Act of 1944, which authorized the construction of a series of dams and reservoirs to protect the cities and farms of North and South Dakota, Iowa, Nebraska, Kansas, and Missouri. The law, also known as the Pick-Sloan Act (after its principal drafters), promised not only to provide flood control, but also to enhance commercial navigation, irrigation, municipal water supplies, hydropower, and recreational opportunities—something for almost everyone.[2] As with most things that sound too good to be true, the 1944 law promised far more than it delivered.

Through the Flood Control Act of 1944, Congress authorized six massive mainstem dams and reservoirs on the upper Missouri in hopes of protecting the population centers and farms of the Missouri River basin and of the Mississippi River below the Missouri's mouth at St. Louis.[3] Congress had an ulterior motive as well, one that was not explicitly stated in the act: Veterans returning from World War II needed jobs (as did Corps engineers), and dam construction was nothing if not labor intensive. In the meantime, Congress began to experiment with yet another reaction to flooding: the provision of federal disaster relief. Through the Disaster Relief Act of 1950, Congress took tentative steps toward developing a coordinated federal response to floods and

Map of the Missouri River basin, showing six mainstem reservoirs
Source: U.S. Army Corps of Engineers

other disasters. The ink was barely dry on the newly minted Disaster Relief Act when the Missouri River jumped its banks again, not just once, but two years in a row. The rain came down in torrents, but the Corps had not yet completed the final dam—Gavins Point, near Yankton, South Dakota.

In July 1951, up to sixteen inches of rain fell on the nation's midsection in only four days. By July 13, over one million acres in Kansas and nine hundred thousand acres in Missouri were knee-deep in water. In Manhattan, Kansas, the business district was submerged under eight feet of water. The river soon swallowed other river towns, including Topeka, Lawrence, and Kansas City. In Topeka, whose name means "to dig good potatoes" in the languages of the Kansa and Ioway Indians, about twenty-four thousand people were evacuated (30 percent of the city's population).[4] Few, if any, potatoes or other crops could be harvested that year.

About an hour's drive west of Topeka, U.S. Army soldiers were enduring boot camp as the 1951 flood swept through and destroyed their barracks. The invasion of South Korea by North Korea just a year earlier, in June 1950, had reignited Fort Riley's historic status as an important military training facility. What had once served as a post for the all-black "Buffalo Soldiers" of the 9th and 10th Regiments of the Army Calvary after the Civil War became a center for thousands of new recruits undergoing basic training and heading overseas.[5] After soggy, grueling days of training, the men were forced to sleep in pup tents that had been pitched on high ground. Although the tents were situated on a ridge, the tent stakes couldn't be anchored firmly in the sodden ground and every once in a while, when torrential rains or wind hit them, the tents would pull up their stakes and float away.

While the soldiers were struggling to keep their cots and their kits dry, flood waters overtopped the levees protecting Kansas City's Argentine and Armourdale areas, and fifteen thousand people were evacuated when water reached the roofs of their houses. As before in the flood of 1903, the Kansas City stockyards were completely inundated in 1951.

The yards' location, situated in the West Bottoms at the confluence of the Kansas and Missouri rivers, had been a boon for commerce and the transportation of livestock, but it was a high risk property. The flood crippled the stockyards' operations, and they never fully recovered their economic viability.

Over in St. Joseph, Missouri, the flood caused the Missouri River to change course, cutting off the city's downtown area from Rosecrans Memorial Airport. In the aftermath, instead of fighting the river back into its original channel, the Corps of Engineers formalized the flood-wrought change by dredging out the new cutoff channel. Even today, travelers must cross the Pony Express Bridge and go through Elwood, Kansas, in order to reach the St. Joseph, Missouri, airport.

According to the American Red Cross, the 1951 flood killed nineteen people and injured over one thousand others. The Weather Bureau reported that thirty thousand farms were affected by high water. The U.S. Geological Survey tallied up the unprecedented economic losses. It concluded, "From the headwaters of the Kansas River to the mouth of the Missouri River at St. Louis, about 2 million acres were flooded, 45,000 homes were damaged or destroyed, and 17 major bridges, some of them weighted with locomotives in an attempt to hold them, were washed away." In all, damages were estimated as high as $2.5 billion —equivalent to a staggering $21 billion loss in 2010 dollars.[6]

The year 1951 was notable for another reason: That fall, thirteen parents of elementary school aged children sued Topeka's Board of Education for unlawful racial discrimination, a case that went all the way up to the U.S. Supreme Court. *Brown v. Board of Education* became the landmark school desegregation case in the country. But in the meantime, another type of racial injustice was unfolding in the Missouri River basin, this time affecting Native Americans. It had nothing to do with school desegregation, and everything to do with the Corps' implementation of the Flood Control Act of 1944, as detailed below.

* * *

THE MISSOURI RIVER was not to remain in its bed after the 1951 flood. The 1952 flood started with snow over the upper reaches of the Missouri River in Montana and the Dakotas. As the snow melted during the warmer spring months, the river swelled once again. Downstream in Omaha, Nebraska, President Harry Truman came to visit. Using the authority provided by the recently passed Disaster Relief Act of 1950, Truman issued an official declaration: Omaha was a federally recognized disaster area. This meant that federal agencies could provide shelter, food, medicine, grants, and other forms of aid to the flood-stricken states, even though state and local governments, as well as non-profit organizations such as the Red Cross, remained on the front lines as the primary responders.

In Omaha, the newly formed Civil Defense team proved itself ready not only for Russian missile strikes (it was the height of the Cold War, after all), but also for floods. Team members and volunteers, joined by around four thousand U.S. troops, covered some fifteen miles of shoreline with flashboards (two-by-four-inch planks) and sandbags, strategically placed to reinforce existing floodwalls. Meanwhile, Union Pacific crews stood watch over the railroad tracks and switchyards on the river's banks. On April 17, 1952, just a few days after Easter, the river crested, setting a record that stood until 2011. Although one thousand acres of industrial land flooded just north of downtown Omaha, the defensive maneuvers were largely successful, and the city withstood the worst of the onslaught. As it continued downstream, however, the Missouri spilled across hundreds of thousands of acres of cropland. Farm levees crumbled, and the river submerged an area up to fifteen miles wide. The flood inundated fourteen hundred houses, half of which were farmsteads.[7]

Remarkably, there were few deaths. As newsman Walter Cronkite reported, "In this record flood, only three lives have been lost so far —and they only indirectly because of the floods. This remarkable safety record is a result of the Weather Bureau's river forecasting service—a fantastic modern weapon against flood dangers. Headquarters receives reports of rainfall, the rapidity of melting snow and other factors—and with a battery of modern gadgets, forecasts just how high the rivers will rise."[8]

* * *

BY THE MID-1960S, half a dozen dams—including three of the twenty largest dams on the face of the earth (Fort Peck, Garrison, and Fort Randall)—were in place on the upper river in Montana and North and South Dakota, and immense reservoirs of water had formed behind them. Since then, the National Research Council estimates that the system has prevented around $414 million in annual flood damages to the lower Missouri basin.[9] But at the same time, it has devastated human and ecological communities.

In particular, the Missouri River project changed forever the lives of Native Americans situated on twenty-three reservations. As historian Michael Lawson points out, it "caused more damage to Indian land than any other public works project in America."[10] The dams submerged 550 square miles—about 350,000 acres—of tribal land. (In total, the dams flooded around one million acres.[11]) Five of the most affected reservations—the Standing Rock, Cheyenne River, Lower Brule, Crow Creek, and Yankton Sioux Reservations—together lost 202,000 acres. A sixth, the Fort Berthold Reservation, home of the Three Affiliated Tribes of Mandan, Arikara, and Hidatsa Indians, lost over 150,000 acres of land —around 20 percent of the reservation's total land base.

This wasn't arid, nonproductive land—far from it. It was the tribes' most fecund bottomland. The reservoirs inundated three-fourths of the reservations' habitat for wild game and nutritious plants and 90 percent of the timberland. When the trees were gone, so were the tribes' lumber, fuel, shade, and windbreaks. Although water supplies were abundant, neither Congress nor the U.S. Army Corps of Engineers took steps to recognize legal water rights for tribes, and even today many of the affected reservations lack adequate drinking water supplies.[12]

When the dams went in, tribal members were forced to relocate their

Chairman Gillette weeps. George Gillette, Sahnish Chairman, and the Business Council
witnessing the sale of 155,000 acres of land for the Garrison Dam and Reservoir,
May 20, 1948

Source: Associated Press (#1393)

homes and their tribal headquarters, unraveling critical communal
ties. In all, the dams and reservoirs displaced more than nine hundred
Indian families. On the Fort Berthold Reservation alone, seventeen
hundred people were forced to move.[13] On the Crow Creek Reservation, Indian families were ousted from their homes, not just once,
but twice. First, they were relocated to make way for the Fort Randall
Dam and the Lake Francis Case Reservoir. With a breathtaking lack of
foresight, the Corps placed these families within the footprint of the
soon-to-be constructed Big Bend Dam. When engineers began working
on that dam just a few years later, the Crow Creek families were moved
once again to make way for Lake Sharpe.[14]

Although the tribes eventually received compensation for their relocation costs and the loss of their land through eminent domain, the

compensation was far less than the tribes had requested, and it was distributed unevenly among the various tribes.[15] Further, the condemnation violated solemn promises that the United States had made to the tribes through the treaties that first established their reservations. Michael Lawson explains that Corps officials were preoccupied with engineering and "had nothing in their training or background that prepared them to deal fairly or knowledgeably with Indians."[16] Tribal access to legislators and competent legal advice was extremely limited and tribal leaders were "simply outgunned." Congress made additional appropriations some fifty years later, but the compensation was still nowhere near satisfactory to the tribes. In 2000, one in every four tribal members residing on the Fort Berthold reservation lived below the poverty rate.

The insult did not end with the dislocation of living members of the tribes. Sacred tribal remains have also been desecrated by federal officials and looters alike. In the Mississippi and Ohio river valleys, only two hundred of over twenty thousand Indian burial mounds have survived intact.[17] Beyond the Mississippi basin, grave robbers in the American Southwest have ransacked between 80 and 90 percent of all ancient burial sites.[18] It is small wonder that funerary objects and sacred items are so attractive to looters: a single funerary basket can bring in up to $150,000.[19] In 1998, sales of Native American art and artifacts at Christie's and Sotheby's auction houses reached an estimated $10 million.[20] Official action has also played a role. For example, remains in the Crow Creek burial grounds and those of other affected tribes have been disinterred and reburied elsewhere to make way for dams and reservoirs.

Missouri River tribes have gone to federal court numerous times to prevent the desecration of their burial grounds. In one case, the Yankton Sioux Tribe wielded a 1990 federal law known as NAGPRA (the Native American Graves Protection and Repatriation Act) against the Corps of Engineers when the operations of the Fort Randall Dam exposed human remains that had been interred at the White Swan Cemetery. Some of the graves in the cemetery dated back to the 1860s, and tribal ancestors had buried their dead in the vicinity since prehistoric times. After the construction of Fort Randall Dam began, a 1950 court order required the United States "to disinter, remove and reinter the bodies of all persons buried [in the cemetery]." The Corps located 428 graves, but according to the court, "failed to effect the removal and reburial of all of the bodies in the cemetery."[21]

In 1965, a deer hunter reported that he had seen human bones near the former cemetery. At that time, reservoir levels had been temporarily lowered pursuant to the flood control operating plan, which called for fall and winter draw-downs to create storage space for subsequent spring season flows from the upstream reservoirs. As it turned out, someone had dug up nearly forty of the graves and scattered the remains on the ground. The Corps removed the remains and reburied them at a new cemetery, but a couple decades later, two local fishermen discovered more bones and parts of caskets along the shoreline near the White Swan site. Instead of reburying these remains, the Corps covered them with heavy white fabric and raised the lake levels above them. In 1999, a park ranger visited the site and observed twenty-five to thirty exposed graves. Human skulls, ribs, hip bones, and femurs were scattered about and caskets rested on the bare ground. The Corps called the tribe's cultural representatives, Ben Gonzalez and Faith Spotted Eagle, "as a courtesy" to inform them about the newest discovery. Gonzalez and Spotted Eagle told Corps officials that they intended to remove and re-inter the remains. The operations manager agreed, but demanded that the tribe provide a removal plan within ten days, on or before Christmas Eve; after that, the water levels would be elevated again and the remains would be flooded. The tribal council voted to file a lawsuit instead.

At the trial, experts testified about a number of options for protecting or removing the remains. The Corps' own archeologist stated that covering the remains with fabric before raising the reservoir's water levels, as the Corps had done in the past, was a terrible idea—the heavy fabric could damage the fragile remains. The court halted the Corps from raising the water levels again until the exposed remains could be removed. Rather than allowing tribal members to collect the remains in accordance with their traditional cultural and religious rites, however, the court told the Corps to do it, and do it quickly, citing the NAGPRA provision that allows the actor to resume its activities (raising the water levels) after thirty days of notification of the interference with burial sites. The court directed the Corps to give custody of the "loose" remains to the tribe, and also ordered the Corps to consult with tribal representatives regarding the protection of remains that were still embedded in the frozen ground, but refused to enjoin the dam operations in the interim.[22] The Corps went ahead and raised the water level ten feet or more, which it believed would protect the remains from looters and from the adverse effects of South Dakota's weather while a long-term plan could be agreed upon. Once again, the tribal ancestors were submerged under the reservoir.

* * *

BY MID-CENTURY, THE federal government had developed an increasingly broad arsenal of weapons to fight the national flood menace. Not only did the country supplement levees with floodways, but it also began to bolster those engineering approaches with informational tools (such as weather forecasting) and economic tools (such as federal disaster relief). But the benefits did not extend equally to all. Some, like the tribes of the Missouri River basin, were forced to sacrifice their centuries-old homeland in an attempt to prevent future harm to downstream cities and farms. Others among the most vulnerable members of our society—poor and minority communities—have experienced the disproportionate effects of dam-building and related flood control efforts as well, as chapter 9 describes.

Just as the 1950s were devoted to containing the Missouri River menace, the 1960s shifted the nation's attention from Midwestern floods to another kind of watery challenge striking downstream in the Mississippi delta: hurricanes.

6

HURRICANE BETSY OF 1965

THE CORPS FORTIFIES NEW ORLEANS AND CONGRESS INSURES FLOODPLAIN RESIDENTS

About five to six thousand years ago, the area where New Orleans now sits began to develop as a sandy mound in the Gulf of Mexico. Sediments from the Pearl River, which forms part of the Mississippi-Louisiana boundary today, flushed out to the sea, gradually building up a barrier chain named the Pine Islands. Over time, as new land built up, the islands became what we know today as the southeastern shore of Lake Pontchartrain. To the west of the Pearl River, sediments from the Mississippi River were also accumulating, but from the opposite direction. About forty-three hundred years ago, the Mississippi began to build up what is known as the "St. Bernard Lobe," which jutted south and east into the gulf. Eventually, this lobe intersected with the Pearl River's westward moving deposits. The circle was completed: Lake Pontchartrain emerged as a distinct waterbody, separated from the gulf by a southern ring of new land—still sandy and soggy from its intimate connection with the rivers and the gulf.

The St. Bernard lobe was not the first delta that the Mississippi River had built. Nor would it be the last. The St. Bernard gave way to

Barrier island formation
Drawing by Professor Stephen A.
Nelson, Tulane University

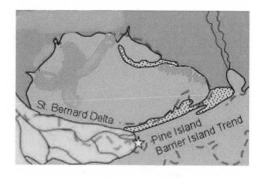

the LaFourche delta, and then to the modern-day Plaquemines delta. Indeed, the Mississippi had been thrashing back and forth for millennia, from west to east, and back again, along an arc about two hundred miles wide. As sediments flushed downstream, they clogged the river's main channel, decreased its depth, and increased its elevation. Always seeking the steepest, straightest path of least resistance, the Mississippi periodically jumped its banks and began to carve new routes to the gulf. Some have likened the river's convulsions to an uncontrolled fire hose. Author John McPhee compared the river's movement to "a pianist playing with one hand—frequently and radically changing course, surging over the left or the right bank to go off in utterly new directions."[1] Metaphors aside, the Mississippi has switched deltas every thousand years or so.

The Mississippi would have changed course yet again in the mid-twentieth century, but for the drastic efforts of the U.S. Army Corps of Engineers. The Atchafalaya River, a major distributary to the west of the Mississippi, draws from the great river at a point about fifty miles above Baton Rouge. By the 1940s, the Atchafalaya River was siphoning off a significant portion of the Mississippi's flow, in volumes equivalent to five Niagara Falls, according to McPhee. Left to its own devices, the Mississippi would have shifted abruptly westward into the channel of the Atchafalaya River. New Orleans would have been isolated from the river that makes it one of the United States' major deep water ports today. The industries and settlements that depend on the Mississippi,

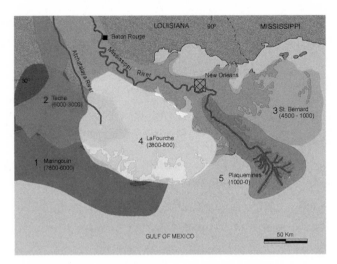

Historic Mississippi River deltas. The numbers represent the
chronological order in which the Mississippi River deltas developed. Present-day New Orleans was once Pine Island.

Drawing by Professor Stephen A. Nelson, Tulane University

from Baton Rouge down to the gulf, would have been cut off from their
lifeblood. Other communities would have suffered the opposite fate:
rather than being abandoned by the old river channel, areas such as
Morgan City that lay in the projected path of the Mississippi's new
channel would likely have been flooded out. In the words of journalist
Thomas A. Lewis,

> If the river succeeded in doing what it had always done, it would leave
> high and dry the Port of New Orleans, devastate the city's economy
> as well as that of Baton Rouge, cut off nearly 20 per cent of the coun-
> try's oil imports and 16 per cent of the nation's fisheries harvest, and
> choke off a major outlet for U.S. agricultural exports. It would leave
> high and dry a chain of refineries and factories stretching from Baton
> Rouge to New Orleans that depend for their existence on the barges
> and the fresh water that the river wants to give to the Atchafalaya. It
> was, well, unthinkable.[2]

In 1954, to prevent such devastating change, Congress authorized the Corps to constrain the Mississippi in its current channel. The Corps cast its mission in terms of a war against the Mississippi, declaring "We are fighting Mother Nature," and "The health of our economy depends on victory."[3] By 1963, the Corps completed a $1 billion series of locks, dams, and floodgates known as the Old River Control Structure. According to congressional mandate, the Corps must operate the structure in such a way that the Atchafalaya captures no more than 30 percent of the Mississippi's flow.[4] Displaying confidence in the Corps' work, officials directed Route 15 right across the top of the structure.

Just ten years after the completion of Old River, the flood of 1973 came perilously close to scouring out the foundation of the mammoth control works. Tons of rocks crashed against the floodgates with the force of a magnitude 8 earthquake. One floodwall gave way. In Lewis' words, "If you stopped a car on the top of the control structure . . . and opened the car door, the vibration of the structure would slam it shut." At the last minute, the Corps completed emergency repairs to prevent the Old River structure from collapsing spectacularly into the Mississippi River. A post-mortem of the badly weakened structure showed just how far the Mississippi had penetrated, just how close disaster had come. McPhee explains the Corps' effort to assess the damage: "What was solid, what was not? What was directly below the gates and the roadway? With a diamond drill, in a central position, they bored the first of many holes in the structure. When they had penetrated to basal levels, they lowered a television camera into the hole. They saw fish."[5]

In 1986, the Corps completed the so-called "Auxiliary Structure" as a backup, designed to prevent the kind of close call experienced in 1973. Still, despite the Corps' interventions, the Mississippi is long overdue for another dramatic change of course. Many believe it is only a matter of time before the Mississippi overcomes its engineered strait-jacket and veers westward down the Atchafalaya channel. In fact, researchers at Louisiana State University's Water Resources Research Institute determined that "It could happen next year, during the next decade, or sometime in the next thirty or forty years. But the final outcome is simply a matter of time and it is only prudent to prepare for it."[6]

* * *

SITUATED JUST SOUTH of Lake Pontchartrain and roughly one hundred miles north of the Gulf of Mexico, New Orleans is surrounded by water. It is hard to imagine a better—or more precarious—place to establish a major settlement. Capturing this ambiguity, geographer Peirce Lewis described New Orleans as an "impossible but inevitable city."[7]

On the one hand, the Mississippi River at New Orleans beckoned early eighteenth-century French settlers as a watery highway for trade with the Indians and with other New World settlers. To the north lay a vast network of rivers and streams—potentially, some fifteen thousand miles of commercial trade routes. And, if the sandbars could be dredged downstream at the river's mouth, then New Orleans could become a port of entry for large ships from abroad.[8] Divine providence had smiled on the French, it seemed, urging them to put down roots in this felicitous spot.

On the other hand, delta marshlands provide a spectacularly unstable foundation for a city. One marine scientist likens the entire Louisiana coastal zone to "stiff yogurt, or soft playdough putty." [9] Although the original settlement claimed the high ground of a natural levee, a big chunk of New Orleans is below sea level—about 45 percent, by some estimates.[10] As described by historian Ari Kelman, water "literally looms above the city, peering down into New Orleans like a voyeuristic neighbor." At the same time, Kelman continues, New Orleans is "basically a floating city," due to its high water table.[11]

Indeed, both impossibly and inevitably, New Orleans has presided over its watery empire since the French settlement of 1718.

* * *

ONE OF THE most comprehensive flood works that affects New Orleans is the Mississippi River and Tributaries Project (MR&T Project), authorized the year after the 1927 flood, and designed to control the levels of flooding experienced in 1927, with an added margin of safety. Despite that massive project, New Orleans' first line of defense is the *back swamp* (marshy areas of the floodplain) of Avoyelles and Concordia parishes. When the back swamp and *batture* (the floodplain between the Mississippi River and its levees) reach their capacity and can store no more floodwater, then the Corps directs the overflow into MR&T floodways and spillways.

Among these, the Bonnet Carré Spillway, almost thirty-three river miles above New Orleans, was designed and constructed in short order after the 1927 flood. Completed in 1931, the spillway pushed the limits of engineering at the time, challenging the Corps to develop concrete strong enough to support the imposing structure. When in operation, the spillway allows the flooding Mississippi to gush into Lake Pontchartrain. But opening the spillway is no small task: it can take up to thirty-six hours for cranes to lift all (or some, as needed) of the seven thousand wooden timbers that stand between the river and the lake.[12] The Bonnet Carré Spillway was activated for the first time in 1937, during the same flood that prompted the Corps to dynamite the Birds Point-New Madrid fuse plugs near Cairo, Illinois. The MR&T project includes at least three additional features that protect New Orleans: the Morganza Floodway, the West Atchafalaya Floodway, and the Lower Atchafalaya Basin Floodway.

The Corps' Immunity

ALTHOUGH THE MR&T pressed the engineers' abilities to their limits, the Corps likely took comfort in the immunity provision of the 1928 Flood Control Act, which shielded it from responsibility "of any kind . . . for any damage from or by floods or flood waters at any place."[13] But despite the simplicity of that statutory language, courts would struggle well into the next century to determine its scope. In general, courts tended to construe the terms "flood" and "flood waters" broadly to apply "to all waters contained in or carried through a federal flood control project for purposes of or related to flood control." As a result, the Corps would be shielded from responsibility for its actions, mistakes, and oversights in a wide range of circumstances.

In a leading case called *United States v. James,* the U.S. Supreme Court in 1986 rejected claims against the government for the deaths of recreational boaters who drowned after being swept through open discharge gates of federal reservoirs that had been constructed for flood control purposes. As attenuated as the relationship between flood control and the creation of dangerous boating conditions may seem, the Court reasoned that the deaths were indeed "related to flood control" within the meaning of the 1928 law.[14]

James left the lower court judges scratching their heads, and in some cases forced them to reach absurd results. Circuit Court Judge Frank Easterbrook put it best:

> [If] . . . "management of a flood control project" includes building roads to reach the beaches and hiring staff to run the project [and if] the Corps of Engineers should allow a walrus-sized pothole to swallow tourists' cars on the way to the beach, or if a tree-trimmer's car should careen through some picnickers, these injuries would be "associated with" flood control. They would occur within the boundaries of the project, and but for the effort to curtail flooding the injuries would not have happened. Yet they would have nothing to do with management of flood waters, and it is hard to conceive that they are "damage from or by floods or flood waters."[15]

The case before Judge Easterbrook, *Fryman v. United States,* involved a teenager who broke his neck when he dove into a submerged hazard in a recreational area managed by the Corps. The teenager—whose injuries made him quadriplegic—claimed that the Corps knew that the hazard existed, and acted not only negligently but also maliciously in not posting warning signs or closing the area to recreation. Compassion aside, Judge Easterbrook felt obligated, under the *James* precedent, to shield the Corps with the cloak of sovereign immunity and to dismiss the victim's claim.

Subsequently, however, the Supreme Court realized that the language in *James* was far too expansive. In the 2001 decision *Central Green Co. v. United States,* a case involving California's Central Valley Project—a project designed for flood control, irrigation, and a variety of other purposes—the Court found that, indeed, the government was legally responsible for damages to a pistachio farm caused by flooding from a federally operated canal. It explained:

> [T]o characterize every drop of water that flows through that immense project as "flood water" simply because flood control is among the purposes served by the project unnecessarily dilutes the language of the statute. The text of the [Flood Control Act] does not include the words "flood control project." Rather, it states that immunity attaches to "any damage from or by floods or flood waters. . . ." Accordingly, the text of the statute directs us to determine the scope of the immunity conferred, *not by the character of the federal project or the purposes it serves, but by the character of the waters that cause the relevant damage and the purposes behind their release.*[16]

In other words, just because the canal, which was built for irrigation purposes, was available to divert water that might otherwise produce a flood, and just because flood control was among the myriad Central Valley Project purposes, it did not follow that immunity should flow with the water that passed through the canal. Therefore, if the damages to the pistachio farm did not result from floodwater but rather from the provision of irrigation water, the government could indeed be held responsible.

* * *

ALTHOUGH THE SCOPE of its legal immunity remained unsettled, the Corps continued to fortify New Orleans against floods, and also labored to secure the city's status as a major deep-water port. New Orleans is tantalizingly close to the Gulf of Mexico, a little over sixty miles as the crow flies, but almost twice that distance as the Mississippi River meanders through shallow bays and coastal marshes. When the initial effort by James Eads to construct jetties (described in chapter 1) no longer sufficed to move modern ships and cargo, the Corps planned the sixty-six-mile-long, thirty-six-foot-deep shortcut known as the Mississippi River-Gulf Outlet ("MRGO" or "Mr. Go," for short). Constructed between 1958 and 1968, MRGO cut off about forty miles of the river's meandering route to the gulf.[17] It proved to be a $95 million boon to navigation.[18]

The engineering challenge was immense. If the floodways failed —Bonnet Carré, Morganza, West and Lower Atchafalaya—the Corps would likely be insulated from liability by the federal immunity

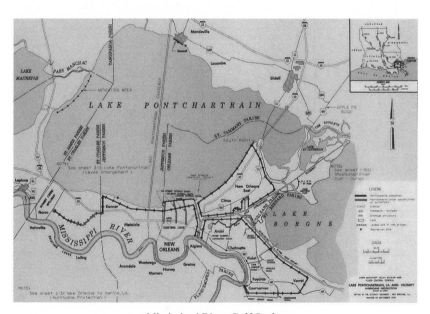

Mississippi River-Gulf Outlet
Source: U.S. Army Corps of Engineers

provision of the 1928 Flood Control Act. But what if MRGO malfunctioned? Would the Corps' flood control immunity extend also to its navigation activities, or would it be held accountable for the damages that followed? In the aftermath of Hurricane Betsy—labeled as the nation's first $1 billion hurricane—courts struggled with that question.

* * *

THE GOVERNOR OF Louisiana was pleased with Louisiana's fortifications on the eve of Hurricane Betsy. Voicing his optimism, Governor John McKeithen testified to a congressional committee, "We have cut the Mississippi in many places so the water can get faster and quicker to the gulf. We have built levees up and down the Mississippi. . . . We feel like now we are almost completely protected."[19] *Almost* completely protected.

Governor McKeithen's confidence was soon shattered. In 1965, Hurricane Betsy whipped across the upper Florida Keys as a category 3 hurricane, then barreled northwestward across the Gulf of Mexico. Betsy made landfall a second time near Grand Isle, Louisiana. Intense winds pushed a storm surge up the Mississippi to New Orleans. Seventy-five people died. Floodwaters submerged tens of thousands of homes in the city and in St. Bernard and Plaquemines parishes, some up to their rooftops. Damages from Betsy were calculated at more than $1.4 billion in 1965 dollars—equivalent to about $8.4 billion in 2000 dollars.[20]

Afterward, a group of flood victims in Orleans and St. Bernard parishes sued the Army Corps of Engineers and claimed that the presence of MRGO had produced unintended, even deadly, consequences. In *Graci v. United States,* the landowners argued that MRGO, then a brand-new marvel of engineering, had done something more than direct barge traffic efficiently to the Gulf of Mexico.[21] Instead, they claimed, MRGO had ushered Betsy's storm surge directly from the gulf to their neighborhoods. The landowners won the first round of their lawsuit, but were roundly defeated in the second. Although ultimately the Corps was vindicated, the lawsuit broke new legal ground in the period before the Supreme Court decided *James*, *Fryman*, and *Central Green Co.*

Importantly, the plaintiffs had at least made it inside the court-house door.

As its first line of defense, the Corps invoked the doctrine of *sovereign immunity*—the idea that the government cannot be forced into court without its consent, deriving from the ancient fiction that "the King can do no wrong." Relying on section 3 of the Flood Control Act of 1928, the Corps argued that it was not responsible for any MRGO-induced flooding because the law provided that "no liability of any kind shall attach to or rest upon the United States for any damage *from or by floods or flood waters* at any place." In essence, so the argument goes, if the Corps undertook to subdue the forces of nature, then it should not be held responsible if it lost an occasional battle in its effort to win the war. The court agreed, but only to a point:

> [T]he purpose of § 3 was to prevent the government from being held liable for the staggering amount of damages caused by natural floods, merely because the Government had embarked upon a vast program of flood control in an effort to alleviate the effect of the floods. Because floods could not be eliminated in a single year, flood damage was bound to recur, and Congress did not want to burden its efforts to lessen the total effect of the floods with the cost of the damage that was certain to result in spite of its efforts.[22]

However, in a critical interpretation of the sparse language of section 3, the court determined that section 3's immunity protects Corps projects *only* if their primary purpose is flood control. MRGO, in contrast, was designed as a navigation project. Therefore, if MRGO had indeed caused the flooding that harmed the plaintiffs, the Corps could not dodge responsibility. As the court explained, "It does not follow that the mere happening of a flood insulates the Government from all damage claims flowing from it."[23]

Despite this initial legal victory, the plaintiffs would gain no satisfaction from the Corps. Harking back to Mrs. Palsgraf's ill-fated lawsuit against the Long Island Railroad Company (recounted in the introduction), the *Graci* litigation determined that the Corps was not the "proximate cause" of the plaintiffs' harm. Siding with the Corps, the court found insufficient evidence to prove that MRGO had caused the landowners' flood damage, or that MRGO had been improperly designed, constructed, or operated. The court observed that the plaintiffs' property had been flooded many times in the past during storms and hurricanes, even before the construction of MRGO. Although the court acknowledged that the plaintiffs suffered far more severe damage during Betsy than during pre-MRGO storms, the court concluded that the "unusually ferocious" intensity and storm-track of Betsy, and not any artificial construction such as MRGO, were responsible for the plaintiffs' damages.[24]

* * *

THIS WOULD NOT be the final word on the Corps' responsibility for failed water projects in New Orleans. After Hurricane Katrina struck in 2005, the courts would have an opportunity to reconsider the issue, as described in chapter 8.

The National Flood Insurance Program and Disaster Relief

THREE YEARS AFTER Hurricane Betsy struck Florida and Louisiana, Congress passed the National Flood Insurance Act of 1968. The notion of a flood insurance program had been introduced over a decade before, in the wake of the 1951 and 1952 floods, when President Truman proposed to set aside up to $50 million for a federally subsidized insurance program.[25] Truman's initial proposal was killed, in part, by the private insurance industry's lobbyists. In 1952, President Truman tried again, this time asking for $1.5 billion for flood insurance to be administered by private industry. It took more than a decade, however, for Congress to provide a meaningful response.

In the interim, in 1956, President Eisenhower floated a proposal for a $3 billion flood insurance program. Eisenhower introduced a new approach by requiring a state-federal partnership to subsidize 40 percent of the premiums. Congress was persuaded to pass the Flood Insurance Act of 1956, but funds were never appropriated for its implementation, in large part due to fears that, rather than limiting losses, the availability of subsidized insurance would cause further development in the floodplains and lead to even greater flood damage.[26] Interest in flood insurance and other risk management tools continued through the late 1950s and 1960s, however, and Hurricane Betsy provided the necessary impetus for the passage of the National Flood Insurance Act of 1968.

The 1968 act established a joint private-government flood insurance program, known as the National Flood Insurance Program (NFIP). In establishing the NFIP, Congress had dual purposes in mind. It hoped that local governments would be motivated to adopt land-use control measures to promote what Congress described as "rational use of the floodplain."[27] In addition, Congress wanted to defray the expense of after-the-fact disaster relief by encouraging floodplain occupants to pay premiums before disaster struck. These goals were to be accomplished through a type of quid pro quo arrangement: the federal government would offer insurance to residents (through private insurers) at below-cost rates, but only if their communities adopted certain land-use regulations and other restrictions. Clearly, Congress intended that the adoption of state and local land-use ordinances prompted by the legislation would serve to reduce flood damage over time.

For an area's residents to qualify for NFIP coverage, the entire community must adopt ordinances to regulate future development in so-called *special flood hazard areas* (SFHAs), which are those areas determined to be within the *100-year floodplain*—defined as an area that has a 1 percent chance of flooding in any given year. The ordinances must meet minimum criteria established by the Federal Emergency Management Administration (FEMA), including zoning restrictions, building requirements, flood-proofing, and emergency preparedness plans. Communities are to develop a series of responses, including the adoption of building codes in the floodplain and construction bans in the immediate floodway. Notably, the NFIP contains an important exemption for properties that already existed at the time the area was identified as a "special flood hazard area," allowing certain older properties to obtain federally subsidized insurance even if the surrounding community failed to regulate future development in the hazard area.

Only five years after the program was enacted, Congress found that flood losses were continuing to increase due to the accelerating development of floodplains. To address the problem, Congress passed the Flood Disaster Protection Act of 1973, which made federal assistance for construction in flood hazard areas—including mortgages and other types of loans from federally insured banks—contingent on the purchase of flood insurance, which is made available only to participating communities.[28] Due to the prevalence of mortgages held by federally insured banks, this amendment remains one of the main hooks that pull participants into the NFIP.

Congress attempted to strengthen the NFIP again in 1988 through the adoption of the Stafford Disaster Relief and Emergency Assistance Amendments. The Stafford Amendments authorize federal funding to acquire destroyed or damaged properties in flood hazard areas, to support rebuilding in non-hazardous areas, and to reduce exposure to flood risk through the imposition of reconstruction standards. Adding another layer of authority to the Disaster Relief Act of 1950, the amendments also allow the president of the United States to deploy federal troops to assist in evacuation efforts and to distribute aid in response to national disasters.[29] But even as Congress adopted laws to encourage communities to adopt land-use regulations that would prohibit new building in flood-prone areas and that would require strengthening of existing structures, the U.S. Supreme Court would decide a series of cases at cross-purposes with those congressional goals.

Two of the most influential of these cases arose out of disputes involving a California floodplain and the South Carolina coastline, respectively. Although telling these tales requires a geographic detour to areas outside the Mississippi River basin, the cases established critical legal precedents that affect every floodplain, wetland, and coastal marsh in the nation, including those in the Mississippi basin.

Camping in the Floodplain and Building in the Dunes

AS CALIFORNIA'S MILL Creek raged out of its channel in 1978, not even the coffins could convince the First English Evangelical Lutheran Church to leave the canyon. Dozens of caskets came bashing down the canyon, unearthed by torrents of water that roared through the Hills of Peace Cemetery. The floodwaters ripped open the coffins, hurling skeletons everywhere. The next morning, mountain residents found human remains in their backyards, a macabre gift of the night of February 10, 1978.

At the time, the church had been preparing its twenty-one-acre property, Lutherglen, for early-season campers. Nestled among the trees of the Angeles National Forest, just above the banks of Mill Creek, the camp provided an inviting respite for church members escaping their suburban lives in Glendale, California, north of Los Angeles. The dining hall, two bunkhouses, caretaker's lodge, and outdoor chapel were clean and stocked, ready to greet a group of handicapped children.

Perhaps it was divine intervention. The flood crashed through Lutherglen on the Friday night before the children were scheduled to arrive. The water destroyed the camp buildings, but the children were still safe at home. For years after, the pastor thanked God. But Reverend Segerhammar was not just praying in gratitude for the church members that had been saved. Nor was he focused solely on mourning the less fortunate souls that had perished on nearby upstream properties during the flood.

The pastor was thankful because his attorney had a plan. They would soon file *First English Evangelical Lutheran Church of Glendale v. County of Los Angeles*, one of the most influential lawsuits ever. As it turned out, they did enjoy remarkable success in court—at first. Despite the initial victory, however, the church's fight would go on for years. And in the end, the church would not recover a dime of compensation from the defendant, Los Angeles County.

After the flood, the church had wanted to rebuild at Lutherglen, but had been thwarted by a temporary flood ordinance passed by the county in the wake of the 1978 disaster. The new law forbade the construction or reconstruction of any buildings within an interim flood protection area along Mill Creek, including the Lutherglen campsite. The church's attorney, Jerrold Fadem, railed against such restrictions. He complained, "The upshot is that the rights of individuals who own property are subject to the whim of local government."[30] Fadem believed the lawsuit raised a matter of great national interest, extending far beyond the Mill Creek floodplain to all of the nation's watersheds. Asserting that the law deprived the church of "all economically viable use" of Lutherglen, Fadem's lawsuit invoked the *regulatory takings* doctrine. That is, Fadem claimed that the law was an unconstitutional taking of property for which the county must pay just compensation under the Fifth Amendment to the U.S. Constitution. Over the next decade, the dispute progressed all the way up to the U.S. Supreme Court.

Releasing its opinion in 1987, the high Court agreed with the church —but only in theory. Chief Justice Rehnquist seized this opportunity to reinvigorate and expand the regulatory takings doctrine—a legal theory generally favorable to landowners and hostile to government regulation and taxpayers. Justice Rehnquist's written opinion, which was joined by five other justices, was hyper-technical, even by the arcane standards of the legal world. Concerned with theoretical legal doctrine, the opinion did not mention that the flood caused the death of ten people and millions of dollars of damage, and that Los Angeles County's new law might prevent similar tragedy in the future. Instead, Justice Rehnquist perceived the church as the victim in this case, suffering from over-zealous government regulations. He worried about floodplain land-owners and expressed concern that the county law might have singled out the church to bear an unfair share of the flood risk in Mill Creek Canyon. Reciting an oft-quoted maxim, Justice Rehnquist admonished that the Fifth Amendment's just compensation provision is "designed to bar Government from forcing some people alone to bear public bur-dens which, in all fairness and justice, should be borne by the public as a whole."[31] Justice Rehnquist concluded that the county would be required to pay the church for the deprivation of its property value, but *only if* the church's version of the facts proved to be correct. In particular, the church would have to prove that the county ordinance had actually deprived Lutherglen of "all economically viable use." That turned out to be a big "if" that posed an insurmountable hurdle to the Church.

The Supreme Court sent the case back to the California Court of Appeal and ordered it to sift through all the facts. (This procedure is known as a *remand.* The U.S. Supreme Court and the state supreme courts decide only legal issues, and often remand the case back down to lower federal or state courts to hold a hearing or trial to determine the facts of the case, and then to apply the articulated legal principles to those facts). In so doing, the California court determined that the church was not entitled to compensation after all. The church had a serious problem: there was no factual evidence that Lutherglen had been rendered valueless by the county ordinance. Even without recon-struction, the court reasoned, the camp property provided ample opportunities for sleeping outdoors in tents and enjoying a wide vari-ety of recreational (and, presumably, spiritual) activities. Further, the court found, the county ordinance was legitimately designed to prevent deaths and injuries similar to those that had occurred during the 1978 flood and during several previous floods. In particular, the debris from

Lutherglen's water-ravaged structures, it said, posed mortal danger to those downstream. If the county allowed the church and other property owners within the floodplain to rebuild, the court continued, those new structures would place life-threatening ammunition in the path of future floodwaters. The court dismissed the church's lawsuit and concluded that, in the end, the evidence proved that the county ordinance had not caused an unconstitutional regulatory taking.

Although the church lost the battle, its lawsuit advanced the war waged by an increasingly active group of property rights advocates who argued that the government must compensate landowners for regulations that interfere with their ability to use their property as they please. After the *First English* saga, nagging legal doubts lingered for decades, leaving government officials uncertain where courts would draw the line under the next set of facts and horrified at what it might cost them to regulate harmful activities. Although the First English Church was not entitled to a financial settlement under the particular facts of that case, the Supreme Court's opinion hinted that the Court might be receptive to future takings claims.

It's hard to know what to make of *First English* and the regulatory takings doctrine. On one hand, it is easy to sympathize with the church and to understand why it wanted to rebuild Lutherglen. The camp enjoys a mile-high panorama of mountains, forest, and water. Certainly Lutherglen could have been rebuilt higher—or at least farther away from Mill Creek—stronger, better. And maybe that would have been enough when the rains came again, perhaps during the night while children lay sleeping at Lutherglen. On the other hand, what if the rebuilt structures weren't safe enough? Perhaps that's not a risk the church should be allowed to push onto others. Despite the Supreme Court's theoretical musings, the final result of the case after remand seems fair: taxpayers should not have to pay the church to follow a law designed to protect its own members and other floodplain occupants from harm.

Despite the county's vindication in this particular case, regulators received the Court's broader message loud and clear. Just a decade after

Congress had created the NFIP in the hope that local governments would restrict floodplain settlement through land-use regulations, the highest court in the land decided *First English* in a way that cast doubt on their efforts. As Justice Stevens worried in his dissenting opinion, "The policy implications of today's decision are obvious and, I fear, far reaching. Cautious local officials and land-use planners may avoid taking any action that might later be challenged and thus give rise to a damages action. Much important regulation will never be enacted, even perhaps in the health and safety area."[32] In fact, just one year later, a group of Louisiana property owners filed a lawsuit claiming that flood control and building ordinances enacted by Plaquemines Parish were unconstitutional under *First English*. Although the federal court in Louisiana ruled against the landowners, the state and federal defendants spent considerable time and money defending themselves. Their efforts to prevent residents from settling in dangerous areas would be vindicated less than two decades later, when Hurricane Katrina made landfall in flood-prone Plaquemines Parish.[33]

* * *

FIVE YEARS AFTER deciding the *First English* lawsuit, the U.S. Supreme Court shifted its focus to a coastal floodplain about three thousand miles east of Lutherglen, and returned to the question of whether the government must compensate landowners when it restricts development on flood-prone property. Like *First English*, this case would directly influence events in the Mississippi River basin, particularly down at the delta.

In *Lucas v. South Carolina Coastal Council*, the Court considered a conflict involving the Isle of Palms, a barrier island off the coast of South Carolina. Sometime before 1986, a developer named David Lucas, together with his business partners, created a posh sixteen-hundred-acre resort on the island. That resort, Wild Dunes, was built as a private gated community that included thousands of homes, two eighteen-hole golf courses, tennis courts, numerous swimming pools, and a spa. The resort proved to be a wonderfully lucrative investment for Lucas and his partners. According to some sources, in its second year alone, the resort generated about $100 million in sales.[34] Over time, the resort also came to enjoy numerous honors, including being named as a top U.S. tennis resort (by *Tennis Magazine*) and as one of the top five hundred hotels in the world (by *Travel & Leisure Magazine*). Meanwhile, back when the resort's development was underway in 1986, Lucas paid $975,000 to purchase the last two lots for his personal use. Two years later, the state passed the South Carolina Beachfront Management Act. Among other things, the law established "setbacks" and prohibited construction within specified distances of coastal dunes and other protected areas. When the law interfered with Lucas' plans to build two houses on his lots, he sued the state. Using a legal theory similar to that advanced by the church in *First English*, Lucas claimed that he should be compensated because the law deprived him of "all economically beneficial use" of his property, and therefore crossed the line into the realm of a regulatory taking.

Lucas conceded that the state law was designed to promote public safety, agreeing in his lawsuit that "discouraging new construction in close proximity to the beach/dune area is necessary to prevent a great public harm," and that an undisturbed beach/dune zone "protects life and property by serving as a storm barrier which dissipates wave energy and contributes to shoreline stability." Nevertheless, Lucas argued that the state should pay him because his oceanfront property was "valueless" in its natural state. The U.S. Supreme Court agreed with Lucas, holding that the state law deprived him of all economic value and therefore constituted a regulatory taking for which compensation must be provided. As a result, South Carolina taxpayers had to reimburse Lucas for his inability to develop his property. The parties negotiated a final settlement of the case, under which South Carolina bought out Lucas for $850,000, with an extra $725,000 for interest, attorney's fees, and costs.[35]

It's easy to understand Lucas' position. Who wouldn't want to build a home along the beach, and who wouldn't be upset if a law prevented one from putting up a new building in an area where numerous beach-front homes already stood? On the other hand, three dissenting justices found South Carolina's side of the story compelling. Like all barrier islands, the Isle of Palms is "notoriously unstable," in the words of dissenting Justice Blackmun. He elaborated: "Between 1957 and 1963, petitioner's [Lucas'] property was under water. Between 1963 and 1973, the shoreline was 100 to 150 feet onto petitioner's property. In 1973, the first line of stable vegetation was about halfway through the property." To protect the island when floods threatened, state and local authorities paid for and completed numerous measures, including sandbagging in the vicinity of threatened structures and undertaking a $1 million beach nourishment project. Justice Blackmun continued: "Between 1981 and 1983, the Isle of Palms issued 12 emergency orders for sand-bagging to protect property in the Wild Dune development [after a state agency determined that habitable structures were in imminent danger of collapse]." In 1989, just three years before the Court decided *Lucas*, Hurricane Hugo struck the Isle of Palms and elsewhere along the South Carolina coast, killing thirty-five people and causing $6 billion in damage.[36]

Following on the heels of *First English, Lucas* reinforced the message that state and local governments adopt protective land-use regulations —including those envisioned by the National Flood Insurance Program —at their peril. As Justice Blackmun fretted in dissent,

> The Court's . . . rule will, I fear, greatly hamper the efforts of local offi-
> cials and planners who must deal with increasingly complex problems
> in land-use and environmental regulation. As this case—in which the
> claims of an *individual* property owner exceed $1 million—well dem-
> onstrates, these officials face both substantial uncertainty because of the
> ad hoc nature of takings law and unacceptable penalties if they guess
> incorrectly about that law.[37]

*   *   *

MEANWHILE, AS THE courts pondered whether MRGO in New Orleans had exacerbated the damage associated with Hurricane Betsy, and as Congress experimented with the notion of federal flood insurance, the Corps industriously set itself to building yet more flood protection measures, including the *Lake Pontchartrain and Vicinity Hurricane Protection Project* (LPVHPP) authorized by Congress in 1965. The project was anticipated to be an $85 million, thirteen-year effort to build control structures, concrete floodwalls, and levees in various areas surrounding Lake Pontchartrain. According to estimates revised in 2005, the cost of construction, recalculated at $738 million, would exceed initial projections almost nine fold. The target completion date was pushed back to 2015.[38] Another hurricane—Katrina—would blast through Louisiana in 2005, long before the project was finished.

But first, the nation's attention turned to its midsection where, in 1993, the national flood insurance program would be put to the test.

7

THE FLOOD OF 1993

REVEALING THE MORAL HAZARD OF
SUBSIDIZED FLOOD INSURANCE

When Congress authorized the creation of the National Flood Insurance Program (NFIP) in 1968, it did so with trepidation. It worried that subsidizing flood insurance—even with strings attached—might encourage development of risky areas, luring even more people into harm's way. Back in the 1920s, President Coolidge had worried that such federal subsidies would encourage waste unless citizens had a direct financial interest at stake. Later, in 1956, Congress established a flood insurance program but declined to fund it because Congress feared that federal subsidies would lead to more floodplain development and increased flood damage. A decade later, renowned geographer Gilbert White, known as the "father of floodplain management," expressed similar concerns. White chaired a federal task force commissioned to reexamine the nation's flood control policies. Its 1966 report supported the concept of federal flood insurance, but with an admonition reminiscent of the concerns voiced by Congress back in 1956: "A flood insurance program is a tool that should be used expertly or not at all. Correctly applied, it could promote wise use of flood plains. Incorrectly applied, it could exacerbate the whole problem of flood losses."[1]

As author Ted Steinberg observes, many may have little choice but to make the floodplain their home, "compelled to live there by economic exigencies, by the simple fact that cheap, flood-prone land can be a magnet for the poverty stricken who are forced to live in the shadow of disaster."[2] But what about other people and businesses that voluntarily choose to occupy (or remain in) the floodplain?

As the private insurance industry knows well, policyholders must be given a financial stake in minimizing their own flood risk to avoid what the industry calls a *moral hazard*. Otherwise, the availability of insurance can prompt people to take risks that they would not otherwise consider, such as occupying river bottomlands highly susceptible to periodic flooding. In 1968, the National Flood Insurance Act finally became law, but generally failed to create sufficient incentives to produce the desired results. Gilbert White's warning would prove to be remarkably accurate in the next great flood.

The Midwest, Submerged

THE SPRING AND summer of 1993 brought torrential rains to the Midwest, along with record-breaking river crests. Throughout the region, the skies dumped a year's worth of precipitation in only three months. The deluge was even more intense in some counties, where twenty inches of rain washed down in a single month. By August, the upper Mississippi and its tributaries, including the Missouri River, flooded seventeen thousand square miles in nine states (an area almost twice the size of New Jersey). According to the National Weather Service, the 1993 flood broke records for both intensity and duration throughout Missouri, Minnesota, Iowa, and Illinois.[3]

The effects of the flood reached far and wide. Levees failed, one after another. The rivers gushed over—or punched through—40 of 226 federal levees. Non-federal levees fared even worse: the torrent over-topped or breached 1,043 out of 1,347 structures. In Missouri, floodwaters sloshed over the steps of the St. Louis Gateway Arch. In

Des Moines, Iowa, the city's water treatment plant was overcome, leaving 250,000 residents without drinking water or sanitation. Throughout the region, submerged Superfund sites, hundreds of discarded barrels, and dislodged propane tanks released their hazardous contents, spreading toxins throughout the floodplain.[4] The U.S. Geological Survey and the Army Corps of Engineers estimated that the flood caused forty-eight deaths and up to $20 billion in property damage. It also displaced about seventy-four thousand people. At the time, the USGS described it as the "costliest flood in the history of the United States."

Many of the victims were uninsured. Although the National Flood Insurance Program had been in place for a quarter century, up to 80 percent of eligible property owners had not taken advantage of the opportunity to protect themselves.[5]

Caskets and Convicts

IN WESTERN MISSOURI, the residents of Hardin watched in horror as the Missouri River, five miles south of town, burst out of its banks, washed over levees, and roared through a local cemetery. The river scoured out about two-thirds of the burial ground, leaving a gash in the earth up to ninety feet deep, by some accounts.

Although previous floods had inflicted damage on other cemeteries throughout the basin, the 1993 flood was remarkable in its desecration of the final resting place of Hardin's ancestors. It unearthed about eight hundred caskets and sent them bashing downstream in the general direction of St. Louis. As the *New York Times* reported, "The remains of whole families floated away, their two-ton burial vaults coming to rest in tree limbs, on highways, along railroad tracks and in beanfields two and three towns away."[6]

A special "Disaster Mortuary Recovery Team"—locally known as the DMORT—retrieved the displaced remains of the city's departed citizens. Volunteers identified the recovered bodies and reburied them. In all, the dead outnumbered the 614 residents of the tiny city that lovingly retrieved and reinterred the caskets.

* * *

WHEREAS THE FLOOD united the citizens of Hardin, Missouri, it unleashed fear and finger-pointing in Quincy, Illinois, after the West Quincy levee failed (on the Missouri side of the river). Released from its strictures, the Mississippi River raged across fifteen thousand acres of farmland and homes, and destroyed more than one hundred buildings. At one gas station, a fuel line ruptured, exploding the station in a spectacular cloud of billowing black smoke and virulent orange flames. Miraculously, no one was killed.

Who was to blame for the West Quincy levee failure: Nature? The Army Corps? According to the Missouri court system, the responsibility lay squarely on the shoulders of one James R. Scott. Scott admits he was on the scene, but claims he was helping to sandbag, not sabotage, the levee. But the state alleged he was up to something far less admirable.

Now known as Inmate No. 1001364 at the Jefferson City Correctional Center, Scott was sentenced to life in prison for violating Missouri's law against "causing catastrophe." That 1977 law made it a Class A felony to "knowingly cause a catastrophe by explosion, fire, flood, collapse of a building, release of poison, radioactive material, bacteria, virus or other dangerous and difficult to confine force or substance."[7] As the Mississippi River threatened to rise above the thirty-foot West Quincy levee, the local levee district strengthened it and the surrounding levees with sand and earth, then lined the levee tops with heavy plastic sheeting to prevent erosion. Volunteers—including James Scott—helped to weigh down the plastic by stacking rows of sandbags on top of it. Despite these efforts, the West Quincy levee ruptured on the evening of July 16, 1993.

Unfortunately for Scott, just after the levee failure, he was one of the people interviewed for a live, on-the-scene television newscast. Some said he even bragged, albeit nervously, about his sandbagging exploits. A local sheriff saw the broadcast and became suspicious because Scott was a well-known troublemaker with a long record of arson and other offenses. After investigation and prosecution, a jury convicted Scott, believing he had removed sandbags from the top of the levee in a deliberate (and successful) attempt to sabotage the structure. That conviction was reversed because the prosecution mishandled evidence at the trial, but a second jury convicted Scott again.

What was Scott's motive? According to the evidence—somewhat muddied because of its improper treatment during the first trial—Scott wanted to strand his wife on the Missouri side of the river where she worked at a truck stop so that he could enjoy an undetected extramarital affair. According to witnesses testifying at trial, Scott boasted that he wanted to destroy the West Quincy levee "to get his wife stuck over there so he could have a party and some stuff."[8]

Although Scott's trials are long over, the verdict is still out in the court of public opinion. Scott maintained his innocence about the levee failure, although after finding religion in prison, he freely admitted to other transgressions. He has at least one outspoken supporter, author and *Time* magazine contributor Adam Pitluk. In his 2007 book, *Damned to Eternity: The Story of the Man Who They Said Caused the Flood,* Pitluk argues that the real culprit was human nature and the instinct to blame a flesh-and-blood person for the 1993 flood, rather than to entertain self-doubts in the face of natural forces. Also to blame, according to Pitluk, is the Army Corps of Engineers for its failure to properly maintain its levees. As Pitluk asserted after his book came out, "The overriding question was whether James Scott caused one levee to fail (even though more than 1,000 levees ultimately failed during the Great Flood), or whether he was merely an easy scapegoat for a community raging at its devastation."[9]

Chesterfield, Missouri: Building and Rebuilding in a Floodplain

CHESTERFIELD SPRAWLS ALONG the Missouri River, just west of its confluence with the Mississippi. A suburb of St. Louis, Chesterfield depended on the 11.5 mile Monarch levee to protect its low-lying incursion into the Missouri River floodplain. The privately built levee sheltered 4,240 acres (about 20 percent of Chesterfield's area) from the level of flooding that has a 1 percent chance of occurring each year (the 100-year flood). But the century old levee finally gave way on July 30, 1993, after it had been pummeled by extraordinary spring and summer rains. The Missouri raged out of its bed and submerged much of the Chesterfield Valley beneath eight feet of muddy water. Chesterfield was one of the hardest-hit communities during the disaster, suffering more than $200 million in property damage and sustaining flood depths of up to fifteen feet. In fact, the flood damage was so extensive that Chesterfield property owners, alone, collected almost 5 percent of the total federal insurance payouts awarded throughout the entire nine-state area affected by the flood.

Chesterfield had begun as a collection of little communities, all shaped by the Missouri River. Originally known as Hog Hollow, the town was founded about 1850. Although no longer known by that name, the initial designation is now honored by the two-lane, hair-pin Hog Hollow Drive that snakes its way down from high land into the flat bottomland of the Missouri floodplain. Another of the original communities, Gumbo, was named for the sticky, oozy, rich river muck that makes the floodplain so fertile for agriculture. A third settlement, historic Monarch, was the namesake of the levee that failed in 1993.

Beginning in the late 1960s, sleepy Chesterfield began to grow. Where it spread into the Missouri River floodplain, it became known as Chesterfield Valley. By the early 1990s, about 240 businesses had sprouted in the floodplain, replacing cornfields and woodland.

Chesterfield had become a thriving, affluent suburb of St. Louis. After the 1993 flood, it would become nationally known for its residents' and business owners' bold exploitation of loopholes in the National Flood Insurance Program.

* * *

IF YOU FIND yourself in St. Louis County—and hungry—you could do a lot worse than to stop at Annie Gunn's Restaurant, one of Chesterfield Valley's most cherished institutions. This local mainstay vibrates with activity, and serves up such delicacies as Grilled Duroc Hog Pork Chop and Sautéed Ozark Forest Mushrooms. If there is a waiting list for seating, as there usually is, you can spend your time in pleasant anticipation by scanning the thirty-two-page wine list. Perhaps you'll waver between a complex chardonnay ("liquid sunlight in the glass") and an elegant pinot noir ("seductively balanced"). Or, you could shrug off convention and choose a playful varietal, such as an Edmeades Zinfandel ("almost like a glass of your favorite cherry pie"). Wine selection completed, you might finish out your wait next door at the Smokehouse Market—the modern incarnation of the 1937 Chesterfield Mercantile Company. Here, you can buy gourmet condiments, olive oils, and cheeses. You can also find anything from pies ("handcrafted" peach pecan) to slabs of bacon ("country cured" and smoked on site).

What can explain the success of Annie Gunn's? St. Louisans are incorrigibly down-to-earth, more likely to order a Budweiser on tap than a sophisticated wine. But their Midwestern souls cannot resist an American success story, and Annie Gunn's provides just that. If it involves grit and adversity, so much the better. And if the tale includes risk-taking and a willingness to work the system to one's own advantage, some people might like it even more.

Back in 1937, the store, tavern, and gas station of the Chesterfield Mercantile Company lay at the end of a dirt road. Just to the north, the half-mile wide Missouri River makes a languorous curve in the final leg of its twenty-five-hundred-mile journey to the Mississippi River.

Nourished by the Missouri's historic overflow, the rich floodplain sustained the cornfields that surrounded the Mercantile. After a succession of shopkeepers, Thom and Jane Sehnert took over the property in 1979. The Sehnerts took pride in the agricultural heritage of the valley, and touted the network of local family farms that supplied them with just-picked produce. In fact, the restaurant adopted the motto, "inspired by the richness of country life." Remnants of that richness appear everywhere, from the surviving cornfields to the nearby street names—River Valley Drive, Greens Bottom Road, Chesterfield Farms Drive. The Monarch-Chesterfield levee system separates Annie Gunn's from the Missouri River. Six lanes of Interstate 64 also lie between the restaurant and the river. But all that paving and engineering cannot change one essential fact: Annie Gunn's lies in the floodplain.

When the Monarch levee failed during the 1993 flood, both Annie Gunn's and the Smokehouse Market were destroyed. The Sehnerts had to be rescued from the rooftop rafters of their business. Otherwise astute business people, the Sehnerts were uninsured until the flood was at their doorstop. According to one newspaper report, the Sehnerts realized on the eve of the flood that their property insurance did not cover flood damage. Just days before the levee crumbled, like many of their Chesterfield neighbors, they bought into the federal flood insurance program. After the flood, with Midwestern gumption, the Sehnerts picked themselves up and began again. The moment was "bittersweet," they explain. It was bitter for obvious reasons. But the flood was also sweet, according to the Sehnerts, because it provided them with an "opportunity"—the chance to rebuild their already-popular restaurant and store into something even better. Seven months and $3 million later, they reopened their business at the same location.

The Sehnerts are celebrities, both nationally and locally. In 2004, the National Association for the Specialty Food Trade named them as one of seven "outstanding retailers" of the year. They also enjoyed recognition from *Gourmet* and *Wine Spectator* magazines. But they are most beloved at home. As the *St. Louis Post-Dispatch* reported, Annie

Gunn's "has woven itself into the hearts of the people who live and work around Chesterfield Valley. Once the most dismal example of the flood's devastation, the restaurant . . . now stands as a symbol of the valley's booming prosperity." This Midwestern impulse to rebuild, to push on in the face of adversity, is undeniably admirable. But is it wise to build, and rebuild, in the floodplain of a major river and in an area that was under many feet of water not too long ago? One can't help but recall the concerns of President Coolidge, the 1956 Congress, and Gilbert White that federally subsidized insurance might create a "moral hazard" that prompts people to take risks they would never consider in the absence of federal subsidies, if only their own money were at stake.

* * *

NOT EVEN THE great Midwestern flood of 1993 could dampen Chesterfield's thirst for growth. In defiance of the Missouri River, Chesterfield, like the Sehnerts, continued to dream big. By 1994, the Army Corps of Engineers had repaired the 100-year-flood Monarch levee. Soon afterward, Chesterfield began a two-front campaign to subdue the river: It lobbied Congress for federal assistance to raise the Monarch levee to the "500-year flood" standard; at the same time, Chesterfield planned to construct its own private 500-year levee if the federal government did not come through. Under a provision of Missouri law related to "tax increment financing," the city designated the floodplain "blighted" and subject to a "redevelopment plan." As a result, Chesterfield acquired legal authority to issue bonds for the proposed levee project, to be repaid with certain sales and property tax revenues. Ultimately, the lobbying paid off. In 2000, Congress authorized the levee improvements as a federal project, paving the way for an estimated $38 million federal contribution to Chesterfield's dream of continued development in the floodplain.

By 2009, Chesterfield Valley supported over 830 businesses—more than triple the pre-flood tally. Most notable is Chesterfield Commons, a 380-acre strip mall that supports such giants as Home Depot, Lowe's,

Chesterfield Commons
Photograph by Christine A. Klein, 2012

Sam's Club, Target, and Wal-Mart. The developer, THF Realty, touts the mile-long development as "possibly the largest, most exciting outdoor shopping center in America." Others simply call it the country's longest strip mall. Just fifteen years earlier, the entire area had been submerged beneath eight feet of water. In the aftermath of the flood, developers spent more than $2 billion on new real estate development.

Today, it is difficult to find much countryside next to Annie Gunn's —the restaurant "inspired by the richness of country life." Instead, Chesterfield Commons is one of the restaurant's closest neighbors.

Accepting Responsibility

EARLIER IN THE century, after the flood of 1927, the nation cried out for federal leadership to control flooding and to help pay for its devastating consequences. In response, Congress developed a three-pronged approach: 1) authorizing construction of more flood control structures, beginning with the Flood Control Act of 1928; 2) providing systematic payment of federal disaster relief, beginning with the Disaster Relief Act of 1950, and 3) providing subsidized federal flood insurance, beginning with the National Flood Insurance Act of 1968. The latter two measures received their first major challenge during the 1993 Midwest flood. Taken together, these reforms proved to be at best incomplete, and at worst utterly wrong-headed. In some cases, they even promoted—rather than relieved—flood damage.

After the 1993 flood, the Clinton administration charged a blue-ribbon committee with the task of studying existing flood control programs and making recommendations for change. The so-called Interagency Floodplain Management Review Committee released its final report in June 1994. Although formally titled *Sharing the Challenge: Floodplain Management into the 21st Century,* the report is perhaps better known as the Galloway Report, named after the committee's executive director, Gerald E. Galloway, brigadier general of the U.S. Army.

Throughout its highly detailed, nearly three-hundred-page report, the committee called repeatedly for shared responsibility and accountability: "All of those who support risky behavior . . . must share . . . in the costs of reducing that risk. . . . Individual citizens must adjust their actions to the risk they face and bear a greater share of the economic costs." The report demonstrated that some of the "lessons" from the 1927 flood had been misinterpreted, resulting in policies that produced unintended adverse consequences. The committee called for several new approaches, including strategic retreat from floodplains, reformation of the NFIP, cutting back on repetitive insurance payments to particularly risky properties, and rethinking disaster relief.

The first suggestion of the Galloway Report—retreating strategically from some flood-prone areas—was rejected by Chesterfield, Missouri, which came roaring back after the 1993 flood. It rebuilt and raised levees, which paved the way for extensive new development in the floodplain. Although lucrative—at least in the short term—such continued reliance on levees may prove to be a risky course of action. The Galloway Report treaded carefully on the topic of levees. Overall, the report concluded that levees "did not cause the 1993 flood" and that "[f]ederally constructed levees, in concert with upstream flood-storage reservoirs, protect many large urban areas from potentially significant damage." However, the report asserted that levees may have adverse "significant localized effects." In addition, the report emphasized the *residual risk* that remains behind levees, as illustrated by the failure of levees designed to protect against the "100-year" flood in Chesterfield, Missouri, and in several other areas. The Galloway Report singled out the Monarch levee as "an example of a levee that induced floodplain development and of the residual risks that result from depending on a levee for flood protection."

Overall, the report recognized that levees encourage a "false sense of security that develops among floodplain occupants." This creates a potentially dangerous cycle. As the report explained, "Reservoirs, like levees, reduce the flood threat to many downstream communities, but the reduction in flood flows simultaneously creates incentives for many people to settle riverbanks and become subject to the impacts of the next major flood. The promise of post-flood support from government and private agencies may encourage people to continue occupying land at frequent risk of flooding."

The committee recognized that communities cannot always build their way out of flood danger with levees, dams, and floodwalls. As a viable alternative, it endorsed floodplain evacuation as a tool to limit unwise development in some cases. According to the committee, using government funds to buy out properties from willing sellers, focusing on those parcels most vulnerable to flooding, would be a viable means of making a strategic retreat from the floodplain.

The buyout approach was not new. In the two previous decades, the government had purchased at least six hundred buildings in the

Mississippi's upper basin and moved them out of harm's way. In 1988, Congress had supplemented state and local efforts with $6 million in federal funds for floodplain buyouts by the Federal Emergency Management Agency.

The 1993 flood stimulated renewed interest in buyouts, and prompted FEMA to make them a key component of its strategy to mitigate flood losses. By year's end, Congress passed the Hazard Mitigation and Relocation Assistance Act of 1993, which made available $130 million to Midwestern communities for disaster relief and hazard mitigation. Recipients were allowed to use funds to elevate buildings, improve drainage, build floodwalls, or take other actions to provide flood protection. Buy-outs became the most popular option, taking nearly 90 percent of the available funds. Although previous buy-out programs applied only in cases where property was repeatedly flooded or where damage exceeded half of the property's market value, the 1993 hazard mitigation program allowed any building in the 100-year floodplain to be bought out. Over two hundred local governments competed for federal funds to acquire buildings in flood-prone areas. As a result, more than ten thousand buildings were removed. Missouri's Community Buyout Program, for example, dedicated more than $30 million in federal money to the acquisition of residential properties. Homeowners received pre-flood value for their homes plus federal loans to find new housing outside flood-prone areas.

Federal funds were also provided for the acquisition of over one million acres of marginal farmlands. Throughout the Midwest, many of those properties were converted to open space, wetlands, and forests. For instance, the Missouri Department of Conservation, the U.S. Fish and Wildlife Service, and the Corps of Engineers acquired tens of thousands of agricultural acres within Missouri's floodplains and converted much of it to wetlands. Likewise, Minnesota spent millions of dollars on conservation easements to prevent development in flood-prone agricultural areas.

Elsewhere, communities adopted ordinances to severely limit new

floodplain construction, but stopped short of removing existing structures and farms. For example, Calhoun County, Illinois—located about
forty miles north of St. Louis—made extensive post-flood revisions to
its zoning code. The revised code prohibited all new residential construction in the 100-year floodplain and required that damaged residences be elevated before they could be replaced. It also limited new
commercial development to river-oriented industries, such as marinas,
resorts, and ferry landings, and required developers of river-oriented
businesses to supply a development application that assured adequate
flood-proofing either by elevating the structures or building a 500-year
private flood levee.

The second suggestion of the Galloway Report called for reform of
several critical flaws in the National Flood Insurance Program. It determined that FEMA's floodplain maps were incomplete or inadequate.
Some counties had not been mapped, particularly in rural areas with
low population densities. Moreover, FEMA policy allowed land to be
removed from floodplain maps if it had been filled with dirt or other
materials to elevate it to a point above the 100-year flood line. The
report's authors feared that such policy "may encourage the filling of
floodplains by developers to avoid community floodplain management
requirements and to assist in marketing flood prone properties. It may
also result in individuals making decisions to purchase a property without full knowledge of the residual risk of flooding, the advisability of
obtaining flood insurance coverage, or access problems during floods."
Owners of parcels thus excluded from FEMA floodplain maps could
avoid purchasing insurance under the NFIP. Overall, FEMA had failed
to map 108 counties that had been declared disaster areas during the
1993 flood.

Another set of NFIP problems uncovered by the Galloway Report
concerned the timing of insurance coverage. When the 1993 flood
struck, NFIP required only a five-day waiting period between the
date of insurance purchase and insurance coverage, which created an
incentive for property owners to wait until flooding appeared imminent

before buying insurance. Annie Gunn's Restaurant was not alone in acquiring coverage just days before disaster struck. More than one-third of the successful insurance claims after the 1993 flood were filed by landowners who had paid into the insurance pool for less than two months. The report estimated that these last-minute insurance policies obtained within sixty days of the flood drained federal insurance coffers by about $105 million. Even more striking, 137 new policy holders waited until just fifteen days before the flood to purchase federal insurance. The majority of these claimants were clustered in Missouri—including Chesterfield—where insurance payouts approached $30 million.

Timing issues also plagued the NFIP program from the opposite direction: Even as some policyholders purchased insurance as late as possible, others terminated their policies as early as possible. That is, although the NFIP directed federally insured lenders to require flood insurance for mortgages attached to floodplain properties, some buyers who had purchased NFIP policies when they first obtained their mortgages dropped coverage when it came time for renewal the next year. As a result of these flaws and other factors, participation in the federal flood insurance program was well below the optimal level to ensure its financially sustainability. At the time of the 1993 flood, a scant 20 percent of insurable structures in the floodplain were covered by federal insurance.

Overall, the report writers worried that the existing NFIP structure subsidized behavior that constitutes a *moral hazard*—a phrase borrowed from the private insurance industry to describe "the situation when an insured party has a lowered incentive to avoid risk because an enhanced level of protection is available." As the report noted, private insurers seek to prevent moral hazards by giving insured parties a financial stake in minimizing the risk against which they are insured. Such measures create financial exposure for policyholders, as through the payment of deductibles, the raising of premiums following the filing of claims, and/or the coverage of only a portion of the policyholder's loss.

The federal program of flood insurance, in contrast, generally lacked such incentives for floodplain occupants to reduce their own exposure to danger.

In 1994, Congress enacted the National Flood Insurance Reform Act to address some of these problems. The amendments took aim at the Chesterfield phenomenon by increasing the waiting period from five to thirty days before newly purchased insurance could take effect. In addition, the 1994 law ratcheted up the pressure on lenders and subjected them to monetary penalties if they failed to enforce the requirement that property owners maintain insurance coverage throughout the loan period. The 1994 law also extended the insurance requirement to all federally *regulated* banks, not just federally *insured* banks.[10]

The Galloway Report's third suggestion called for a comprehensive strategy to address the *repetitive loss* problem—when an NFIP-insured property floods multiple times, and the owner receives multiple NFIP payments. In the nine Midwestern states that flooded in 1993, the NFIP covered a total of 5,723 buildings that had sustained repetitive losses. In the fifteen years before the flood, NFIP had paid 16,978 claims on those buildings—an average of almost three claims per building. Nationwide, repetitive loss properties constituted only about 1 percent of NFIP policies. And yet, this small segment of properties received over 40 percent of all claims payments. The report concluded that such repetitive loss buildings represent a "significant liability" for the NFIP program.

The National Flood Insurance Reform Act of 1994 included some measures to address repetitive losses, including the creation of a Flood Mitigation Assistance (FMA) program of federal grants to assist policyholders in elevating, flood-proofing, demolishing, or relocating structures subject to substantial or repetitive loss. Ten years later, Congress granted FEMA additional authority to strengthen the FMA grant program by providing for penalties against those who declined FMA grants.[11]

For its final recommendation, the Galloway Report focused on federal disaster assistance, which it described as a program "funding disaster." In the five years preceding the 1993 flood, the federal government had spent more than $27.6 billion on such assistance. In many cases, disaster relief was authorized through emergency appropriations that contribute to the federal deficit without the careful planning, fiscal offsets, and continuing oversight that are typically required for nonemergency spending.

As the Galloway Report observed, disaster subsidies can undermine the federal insurance program because some property owners have the false impression that disaster relief provides after-the-fact "free" (or remarkably low cost) compensation comparable to that provided by flood insurance. As a result, such landowners may decline to participate in the NFIP, reasoning that it makes no sense to pay for insurance that might not be needed, if those landowners can count on receiving disaster relief for free if their property floods.

The Galloway Report suggested that disaster relief is not an adequate substitute for insurance, at least from a systemic perspective. The report found that flood insurance is preferable to disaster assistance because the former "internalizes" risk by requiring property owners to pay insurance premiums, albeit at subsidized rates. As a result, the insured landowners may be far more likely to take preventative measures to minimize their flood risk. The Galloway Report acknowledged the deep emotional urge to pay the victims of disaster. After all, it noted, compassion is an understandable (and politically expedient) response to live media coverage of levees bursting and houses swept away by torrents of raging flood waters. But, the report argued, emotion may lead the nation to ask the wrong questions. Instead of wondering how devastating flood damage could happen, it suggested, the better question might be "why the house was there in the first place."

The Galloway Report advocated tough reforms. Most striking was its call to deny full disaster assistance to floodplain occupants who declined to purchase federal flood insurance. According to the report, "If federal response to disaster relief is driven by the immediacy of an event, rather than by rational decision-making, the effort to put everything back to the way it was may increase future risk rather than promote long-term solutions to risk reduction." Instead of indiscriminate disaster relief, which can subsidize bad decisions, the report concluded, the country would be better served by learning from each flood and using it as an opportunity for change.

* * *

JUST AS THE country was absorbing the lessons from the 1993 flood and the Galloway Report, it would be faced with one of its biggest storms yet—Hurricane Katrina of 2005. Our attention would turn from the perils of developing river bottomland to the danger of destroying coastal wetlands, our first line of defense against hurricanes.

8

HURRICANE KATRINA OF 2005

REVEALING THE IMPORTANCE
OF COASTAL WETLANDS

Wetlands, according to the U.S. Geological Survey, are "transitional areas, sandwiched between permanently flooded deepwater environments and well-drained uplands."[1] They come in a wide variety of types, including marshes, swamps, forested wetlands, bogs, wet prairies, vernal pools, and mangroves. They are prominent in natural floodplains. Like floodplains, and despite their name, wetlands are not always wet. But their intimate relationship with water leaves telltale signs—characteristic wetland soils and plants adapted to live in a sometimes-wet, sometimes-dry environment. As a result, scientists can recognize wetlands as such, even during dry periods of the year.

Historically, wetlands have been underappreciated, even disdained. In 1900, the U.S. Supreme Court called for their draining or filling, describing wetlands as "nuisances" and as "the cause of malarial and malignant fevers."[2] Later in the century—much later—some courts began to understand the value of wetlands, describing them as "ecological treasure," [3] "precious resource,"[4] and as "vital to the well-being of both humans and wildlife."[5] Likewise, the U.S. Geological Survey recognizes their value, describing wetlands as "among the most productive habitats on earth" and citing them and their associated aquatic habitats as "critical to fish and wildlife as well as economically and recreationally valuable to humans." In a 1997 article in *Nature*, Robert Costanza (an ecological economist and student of pioneering ecologist Howard T. Odum) and his co-authors famously estimated that the ecological systems of the earth performed services worth at least $16 to $54 *trillion* each year, as compared to a global gross national product of about $18 trillion annually. Wetlands alone, Costanza estimated, contributed an estimated $14.9 billion to the world economy each year by filtering pollutants, recharging groundwater supplies, and providing essential habitat for fish, waterfowl, and wildlife.[6]

Some questioned Costanza's valuation of the world's "ecosystem services" as too high. Still, it gives one pause for thought: from malignant "nuisance" to $14.9 billion treasure in less than a century.

The growing literature on *ecosystem services* seeks to identify, monetize, and quantify the benefits performed by wetlands, floodplains, and other natural areas. Wetlands perform at least five critical functions. First, they provide protection against floods, hurricanes, and shoreline erosion by storing excess waters and releasing them slowly. A Mississippi River basin study found that wetlands destruction and levee building had reduced the basin's natural storage capacity from sixty days of floodwater to twelve days of floodwater.[7] Some experts believe that the height of hurricane-related storm surges can be reduced by one foot for every 2.7 linear miles of coastal wetlands.[8] Even a small

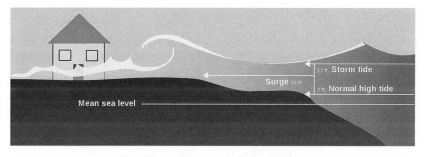

Graphic representation of a storm surge
Source: NOAA National Hurricane Center

reduction in the height of a storm surge translates into a significant degree of protection for coastal communities.

As a second benefit, wetlands improve water quality by processing, decomposing, and trapping inorganic nutrients, organic wastes, and suspended solids that would otherwise pollute surface waters.[9] Site-specific studies have valued this service in the millions of dollars for individual communities. For example, the Environmental Protection Agency (EPA) estimates that one wetland in South Carolina—the Congaree Bottomland Hardwood Swamp—performs water purification services equivalent to those provided by a $5 million treatment facility.

Third, wetlands provide habitat for fish, wildlife, and plants, making them what the EPA describes as "some of the most biologically productive natural ecosystems in the world, comparable to tropical rain forests and coral reefs."[10] Wetland-dependent species contribute approximately $79 billion annually to the nation's $111 billion commercial and recreational fishing industry.[11]

Fourth, floodplains and wetlands help to maintain favorable atmospheric conditions by storing carbon in peat, thus helping to control global warming and climate change. When drained or filled, wetlands release the carbon as carbon dioxide, a greenhouse gas that disrupts the earth's climate.

Finally, floodplains and wetlands provide aesthetic, recreational, and educational opportunities. Studies estimate that Americans spend more than $59 billion annually in connection with wetland-related hunting, fishing, bird watching, and wildlife photography.[12]

The wetlands of the Mississippi River floodplain have been singled out for special recognition. In 2010, nations party to an international treaty—the Ramsar Convention—designated 300,000 acres of the upper Mississippi floodplain as a "Wetland of International Importance." In acknowledging the designation, the U.S. Secretary of the Department of the Interior stated, "The ecological, social, and economic values of the Upper Mississippi River make it one of the crown jewels of this nation's wetlands." Downstream in the Mississippi delta of Louisiana, *coastal wetlands* (also known as *coastal marshes*) make up about 61 percent of the Delta basin.[13] This area has been dubbed "America's Wetland" in a campaign to raise awareness of the value and vulnerability of the area. The campaign explains, "America's Wetland is one of the largest and most productive expanses of coastal wetlands in North America."[14]

Despite the valuable services wetlands perform, the nation has developed, filled, or otherwise lost over half of its original wetlands. In the early 1600s, the area now occupied by the United States (excluding Alaska and Hawaii) included approximately 221 million acres of wetlands. By 2004, that number had declined to about 108 million acres. In the Mississippi River basin states of Illinois, Indiana, Iowa, Missouri, and Ohio, that decline is even steeper: those states have lost 85 percent of their pre-European settlement wetlands.

This downward trend is particularly alarming to the residents of the Mississippi delta region. Each year, portions of the shoreline crumble into the Gulf of Mexico. In part, this process is natural. Deltas compress and sink under their own weight, subsiding downward. But at the same time, sea levels have fluctuated over time. Most recently, the seas have been rising and, as the saltwater intrudes on the sinking coast, it kills freshwater marshland vegetation. As the plants die, the area becomes prone to erosion, allowing the sea to intrude even farther inland.

* * *

ALTHOUGH SOME WETLAND loss can be natural, human activities have greatly accelerated the process. According to the U.S. Geological Survey, human contributions include sediment-trapping dams on the Mississippi River's tributaries (especially on the Missouri River), chemicals used by the agricultural sector (which can impact wetland vegetation), navigation canals (the excavation of which allows saltwater to invade freshwater environments), and the dredging of canals and pipelines to facilitate oil and gas production.[15] The Mississippi River's vast network of levees contributes significantly to the problem. As the

USGS explains, levees can choke off deliveries of freshwater and land-building sediments that might otherwise replenish coastal marshes.

What happens to these sediments, if they are prevented from spreading out laterally into the floodplains or stabilizing the coastal wetlands where the Mississippi pours into the gulf? As writer John McPhee explains, since the early twentieth century, the Mississippi's levees have funneled much of its sediment load and "shot [it] into the Gulf at the rate of three hundred and fifty-six thousand tons a day—shot over the [continental] shelf like peas through a peashooter, and lost to the abyssal plain" of the deep ocean floor. Sediment starvation—combined with erosion and subsidence—sets the stage for coastal catastrophe. As a result of these processes, the Gulf of Mexico has been creeping inward relative to land. By some estimates, New Orleans is a full twenty miles closer to the eroding coast than it was in 1965 when Hurricane Betsy struck, reducing the city's land buffer against hurricanes and floods.

In Louisiana alone, each year the gulf claims an average of twenty-four square miles of coast, an area about the size of Manhattan. This translates roughly into the loss of one football field's worth of land every thirty-eight minutes. It represents the most rapid loss of coastline occurring anywhere in the world. Overall, since the 1930s, the Louisiana coast has lost about twenty-three hundred square miles of dry land, an area equivalent to that occupied by the state of Delaware.[16] If nothing changes, experts project that the gulf will advance about thirty more miles inland by mid-century, leaving New Orleans and other cities vulnerable to the force of the open water.

But disaster would strike long before that, as the area's first line of defense—coastal wetlands—literally crumbled into the gulf.

Washing Away

THE AREA WITHIN a one-hundred-mile radius of the New Orleans has endured forty hurricanes from 1842 through 2008—one every four years, on average. Of those, a dozen were category 3 or greater,

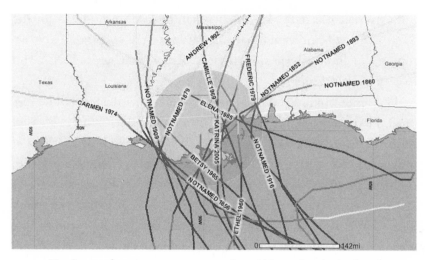

Hurricanes of category 3 or greater passing within one hundred miles of
New Orleans, 1852–2005
Source: NOAA National Hurricane Center

including Hurricane Betsy of 1965. When plotted on a map, the storm
tracks look like so many strands of spaghetti strewn across the region.

In 2002, the New Orleans *Times-Picayune* published "Wash-
ing Away," a series of news articles warning that vast swaths of New
Orleans were vulnerable to hurricane storm surge, despite billions of
dollars spent on levees and other structures to fortify the city. The message of the series was frightening: "Today [2002], billions of dollars'
worth of levees, sea walls, pumping systems and satellite hurricane
tracking provide a comforting safety margin that has saved thousands
of lives. But modern technology and engineering mask an alarming
fact: . . . south Louisiana has been growing more vulnerable to hurricanes, not less."[17]

Co-reporters John McQuaid and Mark Schleifstein were prescient.
Although they wrote a full three years before Hurricanes Katrina and
Rita would ravage the Gulf Coast, their predictions were dead accurate.

The series articulated an awful truth: levees can fail. McQuaid and
Schleifstein revealed that the Corps' confidence in its levees was based

on forty-year-old data. An independent analysis conducted for the *Times-Picayune* indicated that some levees were vulnerable to over-topping, including those in St. Bernard Parish, St. Charles Parish, and eastern New Orleans.

The series predicted that the poor would suffer the most. The segment titled "Left Behind" ticked off the city's obstacles to a successful hurricane evacuation: Low-lying escape routes would be likely to flood well in advance of a hurricane strike. The area's large population (more than one million) would require up to eighty-four hours to navigate eighty miles of vehicle-choked, flooded roads leading to higher ground. Many of the city's poor, including about one hundred thousand people who cannot afford cars, would be forced to rely on what the reporters described as "an untested emergency public transportation system." Citing the lack of sufficient buses, one retired official quoted in the article predicted, "Between the RTA and the school buses, you've got maybe 500 buses, and they hold maybe 40 people each. It ain't going to happen." As a chilling reminder of the immensely high risk, the series reported that to protect workers and evacuees alike, the Red Cross had already "decided that operating shelters south of the Interstate 10-Interstate 12 corridor is too dangerous." (I-10 and I-12 run along the southern and northern perimeter, respectively, of Lake Pontchartrain.) Instead of evacuation, as McQuaid and Schleifstein explained, some parishes would offer shelter in places such as the Superdome or other "refuges of last resort—buildings that are not safe but are safer than homes."

The reporters also recognized that existing flood control and navigation structures pose a serious threat. In a previous article, Schleifstein (with co-reporter Keith Darce) described the Mississippi River-Gulf Outlet (MRGO) as a "hurricane superhighway." Similarly, the 2002 series predicted, "Sinking land and chronic coastal erosion—in part the unintended byproducts of flood-protection efforts—have opened dangerous new avenues for even relatively weak hurricanes and tropical storms to assault areas well inland." Quoting a South Lafourche levee manager, the reporters warned, "There's no doubt about it . . . [the] biggest factor in hurricane risk is land loss."[18]

With the benefit of hindsight, the predictive reporting in the *Times-Picayune* seems nothing short of miraculous. In truth, much of the series' content was widely known, if not widely acknowledged. But McQuaid and Schleifstein had the courage—or the audacity, according to some—to sound the warning loud and clear. It was not an easy message to accept. Unfortunately, the warning would be insufficient to prevent the disaster that would ravage the area just three years later.

Category 3, Magnified

HURRICANE KATRINA WAS one of the most devastating hurricanes that ever struck the United States. As it roared across the Gulf of Mexico, Katrina reached category 5 on the Saffir-Simpson hurricane scale. By the time it made landfall in southern Plaquemines Parish, Louisiana, its fury had diminished to category 3, but its destructive force was far from spent. Katrina raged across a 93,000 square mile area with sustained winds of up to about 125 miles per hour. It triggered an immense storm surge along the northern Gulf Coast—topping out at about twenty-seven feet in some areas. An estimated 80 percent of New Orleans was awash in floodwater up to twenty feet deep. Katrina made history as the deadliest hurricane since 1928, killing directly or indirectly at least 1,800 people in Louisiana, Mississippi, Florida, Georgia, and Alabama. (The 1928 Okeechobee Hurricane killed over four thousand people in Florida, Puerto Rico, and the Caribbean, as recounted in Zora Neale Hurston's 1937 classic novel, *Their Eyes Were Watching God*.) Katrina also entered the record books as the costliest U.S. hurricane. Estimates of the damage varied widely, from about $40 billion to $80 billion of insured losses, and possibly the same amount of uninsured losses.[19]

Although Katrina crossed Louisiana as a category 3 hurricane, its power was magnified by a confluence of factors. As predicted by the *Times-Picayune*, the Mississippi River-Gulf Outlet indeed acted as a "hurricane superhighway." Together with the Gulf Intracoastal Waterway, the two shipping canals funneled the storm surge directly toward New Orleans, energizing and concentrating it into a deadly force.[20]

In addition, the city's system of levees failed miserably, both through over topping as water levels rose above the height of the levees, and through breaching as breaks developed in the floodwalls and in some cases pushed them right over. Authorized in the aftermath of Hurricane Betsy of 1965, the Lake Pontchartrain and Vicinity Hurricane Protection Project (LPVHPP) was an estimated 60 to 90 percent complete.[21] It was supposed to provide security to the city's residents. Instead, it likely lured people into the path of danger with false assurances of safety.

The inadequacy of New Orleans' levees became obvious during Hurricane Katrina and its aftermath. During the initial strike, events unfolded during an agonizing six-hour period. In the pre-dawn hours before Katrina made landfall on August 29, 2005, the Industrial Canal began to leak water into surrounding neighborhoods. By dawn, portions of the Lake Borgne levee began to crumble. At 6:10 a.m., Katrina hit land on the west bank of the Mississippi River in Plaquemines Parish, triggering high winds and a twenty-one-foot storm surge that rose above nearby levees. By 6:30 a.m., the storm surged into the heart of the city through MRGO and the Intracoastal Waterway. The levees began to give way, flooding residential areas of eastern New Orleans. By 6:50 a.m., the funneled surge was pouring over floodwalls and levees into the Upper and Lower 9th Wards, Upper St. Bernard Parish, Gentilly, Bywater, Treme, and Broadmoor. By 7:45 a.m., catastrophic breaches developed in levees along the Industrial Canal. By 10:30 a.m., additional failures occurred on the east and west sides of the London Avenue Canal and at the 17th Street floodwall and levee. Overall, the levee system breached in nearly thirty places, and unleashed floodwaters that continued to rise for several days.

The damage was also magnified by botched or inadequate responses at all levels of government. In the New Orleans metropolitan area many were left stranded as floodwaters rose, just as the *Times-Picayune* series had foretold. Extensive media coverage etched heartbreaking images into the national consciousness—families stranded on rooftops waving flags and flashing signs to attract the attention of helicopter rescuers; forlorn pets separated from their families. Even after reaching designated shelters, victims did not fare well. The New Orleans Superdome, the city's shelter of last resort for some twenty-six thousand evacuees, was poorly run and quickly descended into chaos. City officials had failed to stockpile sufficient fuel, food, water, cots, and medical supplies. To make matters worse, a portion of the shelter's roof was torn off during the storm and the power went out. The toilets stopped working and excrement flowed over the floors. By the time the National Guard evacuated the refugees, piles of trash five feet deep filled the Superdome.

* * *

THE FACE OF Katrina's stranded victims was disproportionately poor and black. Some victims leveled charges of overt racism against officials in the face of painfully slow and inadequate rescue efforts, raising the specter of the denial of evacuation services to black workers and refugees following the Mississippi flood of 1927, or the early nineteenth-century conscription of slaves to build flood-control levees. Even in the absence of intentional discrimination, settlement patterns had long ago stacked the deck against the poor. According to author John McPhee, "Underprivileged people live in the lower elevations, and always have. The rich—by the river—occupy the highest ground. In New Orleans, income and elevation can be correlated on a literally sliding scale: the Garden District on the highest level, Stanley Kowalski in the swamp."[22] (Stanley Kowalski, a character in Tennessee Williams' drama *A Streetcar Named Desire*, is a likeable villain who lived in a low-rent district of New Orleans).

Hurricane Katrina's storm surge poured over floodwalls and levees into the impoverished, mostly black neighborhoods of the Upper and Lower Ninth Wards. Residents without vehicles had no evacuation

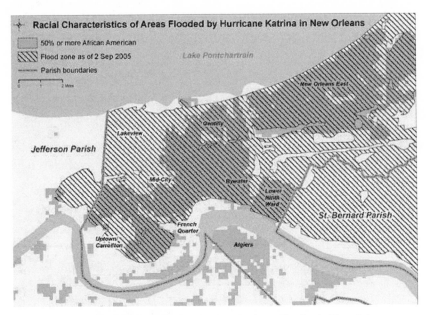

Racial characteristics of areas flooded by Hurricane Katrina in New Orleans

Source: NASA Earthdata, courtesy of Lynn Seirup,
Center for International Earth Science Information Network

options and were left behind, stranded, as the floodwaters rose. In their testimony before a special congressional subcommittee on the role of race and class in the response to Katrina, community activists like Leah Hodges accused officials of overt racism: "If it was not poor African-Americans . . . most affected by this, there would have been a [rescue] plan in place. . . . [These black victims] died from abject neglect."[23]

In the aftermath of the flood, those who stayed in New Orleans were exposed to an array of environmental hazards, ranging from malfunctioning sewage treatment systems to toxic air and contaminated drinking water to excessive levels of formaldehyde in their FEMA-issued trailers. The flood zone in the corridor between New Orleans and Baton Rouge included five Superfund sites and nearly five hundred industrial facilities where a complex brew of dangerous chemicals was stored. During the hurricane, six major oil spills occurred and about sixty underground oil storage tanks were disturbed.[24]

Ethel Williams, a self-described eternal optimist, thought things were looking up when President Bush stopped by her house for a photo-opportunity in April 2006, eight months after Katrina hit. Mrs. Williams, then seventy-four years old, had owned her house on Pauline Street in the Upper Ninth Ward since 1965 and, with her husband, had raised four children there. The hurricane impacted the Williams' home so severely that it had to be gutted after the floodwaters receded. One eventful morning, in preparation for the president's visit, volunteers from Catholic Charities showed up and cleared out the damaged walls, sodden flooring, and other debris. Only the bathtub and a toilet remained in the house. Later that day, President Bush arrived and stood beside Mrs. Williams with his arm draped over her shoulder. While the cameras were rolling, he promised, "We've got a strategy to help the good folks down here rebuild."

Federal funding and lots of volunteer labor were key components of that strategy. National Public Radio reported that Mrs. Williams became "a national symbol of hope that help finally would come."[25] Later, at the president's invitation, Mrs. Williams visited the White House. But when NPR's reporter David Greene went back to New Orleans to check on her four months after the president visited her home, he discovered that little or no progress had occurred. Mrs. Williams' house still stood empty and gutted, just as it was when President Bush stopped by.

Finally, two years after Katrina devastated her house and neighborhood, Mrs. Williams got a check for more than $100,000 to rebuild her home on Pauline Street. The money was from a federal fund, distributed through the state's Road Home program. Mrs. Williams was one of the lucky ones in that respect: out of the 184,000 people in the city who had applied for the money, only 42,000 had gotten their checks by late August 2007.[26] The work on her house was nearly finished by Christmas 2008, but she never got to enjoy her newly rebuilt home. By the time it was ready, Mrs. Williams was too sick to move in. She died of breast cancer in January 2009.[27] For her, disaster relief came far too late.

* * *

IN CONTRAST TO the political effects of the Mississippi River flood of 1927, which helped propel then-Secretary of Commerce Herbert Hoover into the White House following his masterful rescue organization, the 2005 hurricanes contributed to the unmaking of a president and a governor, and nearly brought down a mayor. Many blamed President George W. Bush for the bungled federal disaster response. In an unwarranted compliment to Michael D. Brown, then director of the Federal Emergency Management Agency, President Bush expressed his congratulations in front of the national media, asserting, "Brownie, you're doing a heckuva job." When Brown's work proved undeserving of praise, the president's unfortunate utterance was repeated time and again, serving as fodder for political pundits. Ten days later, Brown resigned from his post. But there was plenty of blame to go around, and the public's discontent with the disaster response was broad enough to reach both political parties, including Louisiana's Democratic governor, Kathleen Blanco, and New Orleans' Democratic mayor, Ray Nagin.

* * *

HURRICANE KATRINA REINFORCED the central lesson of the flood of 1927: when flood control structures fail, they exacerbate greatly the damage caused by natural storms. Following the 1927 flood, federal officials had determined that the Army Corps of Engineers' approach to flood control had been a "monumental blunder," causing a disaster that was "man-made" rather than natural. Almost eight decades later, analysis of the 2005 hurricane season yielded a similar conclusion: much of the damage was of human, rather than divine, origin. Refusing to attribute the Katrina damage to an "act of God," the *Wall Street Journal* quipped, "God is getting a bum rap." In fact, the *Journal* noted that according to some climate scientists, the storms themselves "could be at

least partly man-made," as warming oceans and rising seas present conditions conducive to hurricane formation, intensity, and storm surge.[28]

Ultimately, the Army Corps of Engineers accepted a large dose of
responsibility for the failed levees and floodwalls. The chief of the
Corps led an interagency task force made up of leading engineers, scientists, and academics. The task force criticized the "system" of federal
and local structures designed to protect the region, describing it as a
"system in name only" that "did not perform as a system." Instead of
a coordinated system, there was only a slapdash mix of floodwalls and
levees, rising to varying heights and constructed of mismatched, poorly
interfacing materials. Without the failure of this faulty assemblage of
levees, the report concluded, up to half of the direct losses from Hurricane Katrina could have been avoided. A series of post-Katrina investigations faulted the Corps for additional errors. Among other things,
the Corps had failed to account for the gradual sinking of native soils
in its floodwall design, had neglected to keep levees and floodgates
in good repair, and had provided insufficient assurance that drainage
pumps would continue to function during a catastrophic storm.

In sum, human error led to unnecessary flooding that transformed
Hurricane Katrina into what the report described as a "catastrophe
within a catastrophe."[29]

Dancing the Louisiana Two-Step

THE LOUISIANA TWO-STEP is a familiar staple to aficionados of
Cajun and Zydeco music. The dance is mostly a mix of quick half-
steps—often two in a row by the same foot. A typical sequence might
be left (quick, quick), right (quick), left (quick), right (quick, quick), left
(slow), right (slow). Although experienced dancers glide together in a
common direction, a novice can get lost in the fancy footwork and lose
track of which side of the room to dance toward.

After Hurricane Katrina, a deluge of lawsuits called on judges to sort
out responsibility for the catastrophe. At times, it seemed as though the

judges were dancing the Louisiana Two-Step, shuffling in directions unpredictable to most observers. But one can hardly fault the judges. The factual bases of the hurricane lawsuits were so confusing that they could make even a hard-boiled lawyer misty-eyed with nostalgia for the simpler days of Mrs. Palsgraf's 1928 case against the Long Island Railroad Company (as described in the introduction). The sequence of cause and effect was challenging enough in Mrs. Palsgraf's case—a railroad employee shoved a passenger, who dropped a package, which exploded, which knocked over a distant scale, which struck poor Mrs. Palsgraf. But at least the Palsgraf events were all visible to the naked eye. Today's judges, in contrast, must sort out a mind-boggling array of cause-and-effect factors at the edge of our scientific understanding.

* * *

AFTER HURRICANE BETSY of 1965, a group of plaintiffs had sued the federal government, claiming that the Mississippi River-Gulf Outlet (MRGO) had funneled storm surge toward the city and caused the flood damage that they suffered. The court ruled against the flood victims in *Graci v. United States*, citing insufficient evidence to prove that the construction or operation of MRGO had been to blame (as described in chapter 6).

Forty years later, Hurricane Katrina devastated New Orleans and other portions of the Gulf Coast. Again, flood victims sued the federal government, alleging that their damages had been caused, in part, by MRGO and by the still-incomplete Lake Pontchartrain and Vicinity Hurricane Protection Project (LPVHPP). As with the Hurricane Betsy litigation, the dichotomy between flood control and navigation proved important. With respect to the former, the court held that the federal government could *not* be sued for failed flood control structures such as the LPVHPP because of its protective immunity.[30]

But this time, something was different, at least in the early stages of the lawsuit. Although immunity shielded the Corps from responsibility for flood control measures gone awry (such as any damages

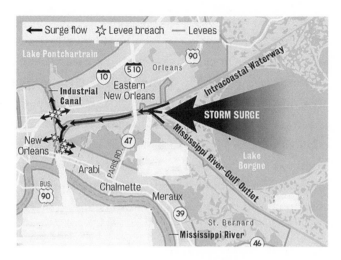

Storm surge during Hurricane Katrina
(graphic by Emmett Mayer III)
Source: (New Orleans) Times-Picayune

caused by the LPVHPP), the court determined that the Corps could
be held accountable if the plaintiffs could prove that MRGO, a *navigation* channel, had intensified Katrina's storm surge and aimed it straight
into New Orleans.[31] This logic followed directly from *Central Green
Co. v. United States* (discussed in chapter 6), a 2001 decision in which
the U.S. Supreme Court dramatically reduced the scope of government
immunity for flood damage, at least where the government controlled
the waters for a purpose other than flood control (there, for the irrigation of pistachios).

In addition, forty years had passed since the Betsy litigation, and
scientists had learned a lot more about the cause-and-effect of hurricane
damage. Wetlands, once dismissed as swamps worthy only of plowing
under, had become valued for their ecological diversity and for the ecosystem services they provide. And one of those key ecosystem services,
we had learned, is the dampening of hurricanes and storm surge.

Federal Judge Stanwood Duval, Jr., rolled up his sleeves and got
serious about educating himself as to the relationship between MRGO,

wetland destruction, and storm surge damage. He spent two years sift-
ing through various legal motions. Judge Duval then presided over a
nineteen-day trial, during which he received technical testimony from
numerous witnesses. The media watched carefully. More than one
reporter described the lawsuit as "groundbreaking" and noted that the
plaintiffs' success could open the door to new claims against the federal government totaling billions of dollars. Finally, the trial was over,
and Judge Duval made his decision. His written opinion occupied
ninety-three pages in the official reporter. Although Judge Duval did
not agree with all of the plaintiffs' contentions, he agreed with most
of them.

In a momentous decision, Judge Duval ordered the Corps to pay
about $700,000 to compensate the victims for the damages they suf-
fered. In the end, the judge decided that the Corps' "monumental negligence" in maintaining and operating MRGO—a *navigational* channel,
not a flood control canal—was a substantial cause of the catastrophic
flooding of portions of St. Bernard Parish and the Lower Ninth Ward.
As in *Graci*, the Corps would not enjoy immunity. But this time, the
court also found that MRGO was to blame. The judge detailed at least
three chains of events that led to the disaster.

First, the evidence persuaded the court that the construction of
MRGO had decimated vast expanses of hurricane-taming wetlands—
about twenty-three square miles in all. In addition to the direct destruction of wetlands during MRGO's construction, MRGO had channeled
the gulf's saline flows into freshwater or low salinity marshes. This, in
turn, stimulated the conversion of existing plants to more salt-tolerant
vegetation, which is more vulnerable to erosion. As Judge Duval
explained, marshland vegetation supplies friction that slows the progress of storm surges. According to the Corps' own measurements, 2.7
linear miles of coastal marsh can reduce the height of storm surge by
one foot. Second, the court determined that the presence of MRGO
accelerated the sinking of adjacent levees at rates three- to five-fold

faster than the subsidence rates of unleveed banks. In one case, a levee adjacent to MRGO sank down seven feet between 1985 and 2005. Finally, the judge determined that the Corps had known that MRGO was experiencing severe erosion, but had declined to armor its banks or otherwise protect them from deterioration. As a result, the channel had widened at an alarming rate. In the Corps' own words (drawn from a 1988 report), "Because erosion is steadily widening the MR-GO, the east bank along Lake Borgne is dangerously close to being breached."

Judge Duval lambasted the Corps and described its attitude as "callous," "myopic," and "negligent." The judge contended that the Corps had done nothing to forestall catastrophe, even "with the knowledge that the erosion problem was potentially cataclysmic for the lives and property of those who lived in St. Bernard Parish." Instead, the Corps had elevated the needs of the shipping industry over all else. In Judge Duval's words, "the sole focus of the Corps was to guarantee the navigability of the channel without regard to the safety of the inhabitants of the area or to the environment."[32] The judge's opinion was affirmed by Judge Jerry E. Smith, writing for a three-judge panel of the Fifth Circuit Court of Appeals.[33]

The importance of Judge Duval's decision, and the Fifth Circuit's affirmance, cannot be overstated. At long last, the federal judiciary had established in unmistakable terms that a significant portion of the past century's disastrous flooding was of human origin—unnatural disasters that were the inevitable consequence of transforming the Mississippi River from a natural river into an engineered shipping and drainage channel.

Even Congress withdrew its support for the Corps' design and operation of MRGO. In fact, in 2006, Congress ordered the Corps to develop a plan for the de-authorization and closure of MRGO from the Gulf Intracoastal Waterway down to the Gulf of Mexico.[34] By June 5, 2008, the de-authorization was official. Thus, while the Corps' lawyers were busy defending legal charges related to MRGO in front of Judge

Duval, the Corps' engineers were beginning to close off the channel with almost half a million tons of rock.

* * *

BUT THAT CRITICAL acknowledgment would be short-lived. Inexplicably, the appeals court abruptly changed course, shuffling in the opposite direction, as if dancing the Louisiana Two-Step. In fact, the appellate court doubled back to a broad interpretation of immunity that the Supreme Court had rejected more than a decade previously. The discredited 1986 opinion, *United States v. James,* discussed in chapter 6, had relieved the Corps of responsibility for recreational boaters who had drowned due to the Corps' negligence, simply because the dangerous reservoir stored flood waters from time to time. Subsequently, under its 2001 decision in *Central Green Co.,* the high Court made clear that *James'* interpretation of immunity had been too broad, and that the Corps' flood control immunity should be much narrower.

Despite such Supreme Court guidance, just six months after writing his first opinion, Judge Jerry E. Smith changed course completely. Writing for the same three-judge panel, he substituted a second opinion in September 2012. In the new opinion, Judge Smith continued to acknowledge that MRGO's design and operation "greatly aggravated" Hurricane Katrina's impact. He also reiterated that the 1928 Flood Control Act's immunity provision for flood control efforts did not protect the Corps for its negligence in connection with MRGO (as mandated by *Central Green Co.*). But the judge determined that an exception to a *different* federal law—the Federal Tort Claims Act (FTCA)—protected the Corps from responsibility for "discretionary" functions, such as the decision not to stabilize the riverbanks supporting MRGO, primarily to save money. Although his opinion may have been in conformity with the letter of the law, it was not consistent with the spirit of *Central Green Co.*

In a statement with potentially vast implications, Judge Smith determined that the FTCA exception completely insulated the government from liability, in order "to prevent judicial second-guessing of legislative and administrative decisions grounded in social, economic, and political policy." No doubt Judge Smith was also influenced by the fact that hundreds of lawsuits had been consolidated into the massive conglomeration of claims before him. Any other decision would have subjected the federal government to potentially ruinous liability.

Impossible City, Inevitable City

THE DEVASTATION FROM Katrina led some to wonder whether New Orleans might just be an "impossible city" after all, in the terminology of geographer Peirce Lewis.[35] Some have suggested, albeit reluctantly, that perhaps it is time to give up, or at least to retreat from, the city's most vulnerable areas.

Still, many hold the view that New Orleans is an "inevitable city" (again, using Lewis' terminology) and that we should do what it takes to help it rebound after each disaster. President George W. Bush reinforced the inevitability view when he toured the city after Katrina. Just weeks after the disaster, President Bush pledged, "This is our vision for the future, in this city and beyond: We'll not just rebuild, we'll build higher and better."

In his book *A River and Its City*, history professor Ari Kelman takes both points of view seriously. He poses (and answers) the question whether or not New Orleans should be rebuilt, despite its low-lying, calamity-prone location. Kelman's answer? A "resounding yes," despite his acknowledgment "that the next disaster is not a matter of if but when."[36] Kelman cites social, cultural, and geographical reasons why the city's "multiracial, multiethnic metropolis" should be sustained. And, he observes, New Orleans is a revenue-generating port city, ranking third in the world before Katrina struck.

Impossible city? Inevitable city? In the end, it is likely that New Orleans is both.

Actor Brad Pitt seems to grasp this dichotomy intuitively. He founded the *Make It Right Foundation* to provide some 150 economically affordable, environmentally sustainable houses in the Lower Ninth Ward neighborhood that was wiped out by Hurricane Katrina. His goal was to provide shelter to lure displaced residents back home and to revitalize the neighborhood. The architecture is strikingly bold and modern, blended with hints of old New Orleans. Some welcomed the style as appealing and innovative, while others found it jarring. Architectural tastes aside, the structures were built to endure in New Orleans, the inevitable city.

But Pitt hedged his bets with designs that gave a nod to the city's impossible, flood-prone location. His builders elevated many of the houses on sturdy pylons to protect them from future floods. Other houses in Pitt's neighborhood are attached to their foundations by twelve foot tethers, ready to float atop rising floodwaters. Still other homes have emergency access to their roofs. There, secure compartments hold an unusual amenity likely found in no other American housing development: a fully equipped life raft.

* * *

TIME AND AGAIN, hurricanes and floods in the Mississippi basin have brought suffering to many, but the poor and people of color often suffer disproportionately. As Kelman suggests, there's a lot we can learn from New Orleans about the relationship among poverty, racial inequities, and "most of the socially constructed disasters that we persist in mislabeling *natural.*" The next chapter takes up the topic of environmental justice.

9

RUINED LIVES

TROUBLE RAINS DOWN ON
MINORITIES AND THE POOR

In Richard Wright's story "Down by the Riverside," which takes place during the Mississippi River flood of 1927, the character of Brother Mann finds himself in desperate need of transportation. Lulu, his pregnant wife, has to get medical attention, and quickly. But river water is lapping at the steps of Mann's house, and it is impossible to walk or drive to the Red Cross hospital. As Mann rows his family past the levee in a stolen boat (there were none for sale, at least not to a black man), he observes "long, black lines of men weaving snake-fashion about the levee-top," wearily stacking heavy bags of sand and cement beside the river. These are not volunteers; white soldiers stand behind them with rifles. When the levee breaks, the "lines of men merge . . . into one whirling black mass" in the raging Mississippi River.[1] Miraculously, Mann makes it to the hospital, but it is too late. Lulu is dead. And Mann himself is compelled to join the sand-bagging effort. In the end, Mann dies, too, shot in the back by soldiers as he tries to escape.

Mann and Lulu are fictional characters, but their story rings true. The treatment of African Americans and the poor during the 1927 flood, described in chapter 3, was no anomaly. In fact, a review of catastrophic floods throughout American history reveals a pattern of

African American flood refugees stand in line at Birdsong Camp in Cleveland, Mississippi, April 29, 1927

Source: 1927 Flood Photograph Collection, Mississippi Archives, Department of Archives & History, Jackson, MS

discrimination against minorities and low-income communities over time and space.

African Americans weren't the only ones who fared poorly under the nation's flood policy. Native Americans, too, bore more than their share of hardship. As detailed in chapter 5, when the dams of the Missouri River flood control project submerged tribal lands, the new property to which the tribes were relocated was far from ideal. According to a tribal-national partnership for environmental education, whose work is supported by a National Science Foundation grant, "Indian reservations are home to some of the most polluted and environmentally degraded sites in the country. Reservations contain a disproportionate share of superfund sites . . . and toxic military sites. . . . People living on reservations have some of the highest incidences of environmentally-related health problems." The report concludes that Indian reservations, which are "the most geographically, economically, and educationally isolated areas in the nation, . . . are least able to cope with the complex environmental challenges that they face."[2]

Environmental Injustice

THE MOVEMENT AGAINST environmental injustice began outside the Mississippi River basin. But as the seven Mississippi flood stories of this book illustrate, the pernicious effects of the injustice are felt throughout the country.

The United Church of Christ (UCC) Commission for Racial Justice brought environmental justice to the forefront in 1987 when it released its landmark report, *Toxic Wastes and Race in the United States of America*.[3] Co-authored by civil rights activist Reverend Benjamin Chavis, the report documented how populations that are marginalized due to race, poverty, language, or cultural differences bear a disproportionate share of society's environmental burdens.[4]

The UCC report was inspired by events that transpired in Afton, North Carolina, when Afton's mostly African American residents took a stand against a 142-acre toxic waste dump. The Afton residents were living next to 81,500 tons of polychlorinated biphenyls, better known as PCBs. PCBs were commonly used in electrical equipment due to their non-flammability, stability, and insulating properties. But PCBs cause cancer, as well as harmful neurological, reproductive, and developmental effects. They are also highly migratory, moving easily through soil and water. The water table near Afton is shallow (only five to ten feet below the surface), making the local drinking water wells vulnerable to contamination.

The PCBs came from the Ward Transformer Company, which rebuilds and sells high voltage transformers. The company representative, Robert "Buck" Ward, hired his long-time friend Robert Burns to remove PCB-laced soil from the facilities in Raleigh, North Carolina. The problem, in Buck's mind, was a new federal law called the Toxic Substances Control Act of 1976, which banned the sale of PCBs and required the disposal of PCBs in carefully controlled landfills.[5] Rather than disposing of the contaminated soil lawfully (and expensively), Burns and his sons loaded it into a tanker truck, drove off, and surreptitiously released it from the tanker's bottom valve along hundreds of miles of roads in fourteen counties throughout the state. After this midnight dumping was discovered in 1982, then-Governor Jim

Hunt ordered the contaminated soil scraped up and moved to Afton, sixty miles away from Raleigh.[6]

Understandably, the residents of Afton did not want to live next to a mountain of PCBs. Over the course of the next six weeks, as the state transported over six thousand truckloads of PCB-laden dirt to "Hunt's Dump," the residents and their supporters marched, carried signs, and lay down in front of the trucks. Two hundred state patrol officers and a unit of the National Guard were deployed for fear that the protests would turn violent. Officials arrested over five hundred participants, including Reverend Chavis.[7]

The NAACP went to federal court, alleging that the state had engaged in discriminatory decision-making in violation of Title VI of the Civil Rights Act of 1964. Congress had enacted the Civil Rights Act in response to the movement led by Dr. Martin Luther King, Jr. and other activists and community leaders. Under Title VI, federal agencies are prohibited from giving federal funding to agencies or programs that discriminate on the basis of race.[8] North Carolina came within the purview of Title VI because the state was a recipient of U.S. EPA funds.

The NAACP argued that the state had discriminated against minorities by building the landfill in Warren County, which had the highest percentage of minorities among all the counties in the state, while ignoring several other suitable or even superior sites in other locations with lower percentages of minorities. The court rejected the claims summarily, stating that there was "not one shred of evidence that race has at any time been a motivating factor in any decision taken by any official."[9] The judge indicated that the Title VI claims were insincere, because the NAACP had not raised racial discrimination as an issue during earlier administrative proceedings or in previous lawsuits related to the PCB dump.[10] To win a Title VI claim, plaintiffs must prove that an action had more than a disparate *impact* on minorities; they must show outright, *intentional* discrimination, which is a very difficult thing to demonstrate in court.[11] After all, in this day and age, no government official is going to admit publicly that he or she decided to place toxic substances in a particular community *because the residents are black.*

It's true that no one wants a dump in the backyard. There's a term for this sentiment: "NIMBY"—Not in my backyard.[12] So why care about Afton? As pioneering environmental justice scholar and activist Dr. Robert Bullard explains, the area "exhibits the 'quadruple whammy' —in that it is mostly black, poor, rural, and politically powerless."[13]

It took almost two decades for the Afton residents to get any type of relief, but they never gave up, and eventually officials removed the contaminated soil from Hunt's Dump and incinerated it. The midnight dumpers were prosecuted for unlawfully discarding toxic substances. Burns ended up serving a five-year jail sentence; his sons, who helped their father spew PCBs all over North Carolina, got off with suspended sentences in return for their testimony against the Ward Transformer Company. As for Buck Ward himself, he was convicted, but released after only nine months in prison. Buck went back to work selling transformers, and he's been ordered to pay for a small part (10 to 15 percent) of the $17.1 million cleanup. Taxpayers covered the lion's share of the costs. Meanwhile, as Dr. Bullard explains, "the innocent community received the onerous 21-year sentence of living in a toxic-waste prison."[14]

Hunt's Dump still exists. The treated soil was placed in a huge pit, graded, and seeded with grass. Despite the greenery, Dr. Bullard notes that the landfill is "one of the most recognized landmarks in the county."[15] The government says the area is completely harmless now, and performs periodic soil, water, and air sampling and testing to ensure that the remaining PCBs are staying put.[16] Warren County's twenty-year land-use plan, "Warren County-2022," identifies recreational use as a future goal for the site. The state has supported these plans by transferring title to the landfill to the county and by removing previously imposed development restrictions. Although the state claims that the site will be just fine for baseball or soccer fields, it recommends against the construction of a swimming pool or any other facility that would entail digging. Apparently, North Carolina believes that, even after all this time, the remaining contaminants are best left undisturbed lest they seep through the soil and groundwater and into the adjoining neighborhood.[17]

Robert Bullard isn't exaggerating when he says that the story of Hunt's Dump has had lasting nationwide effects. Not only did it energize the environmental justice movement, but the landfill and the discovery of other toxic dumps like the Love Canal neighborhood in upstate New York and the PCB-ridden Valley of the Drums hidden away in a Kentucky forest motivated Congress to pass the Comprehensive Environmental Response, Compensation, and Liability Act (better known as the Superfund law). The statute, enacted in 1980, authorizes the U.S. EPA to identify and remediate illegally dumped hazardous substances that threaten human health or the environment. In 2010 —the thirtieth anniversary of the law—the agency boasted that it had "completed construction of cleanup remedies at 67.5 percent of final and deleted sites on the National Priorities List . . . [and] readied nearly 1.3 million acres of land for return to productive use." Funding shortfalls, however, have limited the EPA's ability to clean up (or, ideally, to force the polluters to clean up) all of the contaminated sites. Congress has refused to renew the tax on chemical and oil companies that had provided money for cleanups—the so-called *Superfund*—thereby leaving the federal taxpayer to pick up the tab for most federally operated cleanups. Critics point out that, somehow, Superfund sites near wealthy white communities still tend to get cleaned up, while sites near poor minority neighborhoods languish.[18]

* * *

HAZARDOUS LANDFILLS AND midnight dumping aside, do minority communities experience disproportionate exposure to damages from floods, hurricanes, tsunamis, drought, and other hazards? The Sioux tribes' experience with the Missouri River project is one example of the disparate impact of flood control policies. The low-income and minority communities of the Lower Ninth Ward, St. Bernard Parish, and other areas of the gulf coast represent another case in point. Not only have they been flooded repeatedly, but in the aftermath of Hurricane Katrina, they have been exposed to contaminated water, leaking fuel tanks, soils laden with toxic chemicals, and building debris pockmarked with asbestos, lead, and dangerous molds.

Susan Cutter, the Carolina Distinguished Professor of Geography and Director of the Hazards Research Lab at the University of South Carolina, explains that these are far from isolated examples: "Environmental injustice in natural disaster incidents occur[s] in a multitude of ways, including the manner of evacuation, property affected by water/winds due to location or pre-disaster mitigation, and in the way reconstruction has (or has not) proceeded."[19] For example, following Hurricane Frederic, which made landfall at Dauphin Island, Alabama, on September 12, 1979, eleven counties in Alabama, sixteen in Mississippi, and five in Florida were declared disaster areas, eligible for federal disaster aid. Frederic was the Federal Emergency Management Agency's first major test; the hurricane hit only three months after FEMA was established. In one sense, FEMA failed miserably. Response workers restored power in black areas only after it was restored in white areas, and black communities received less emergency shelter, ice, food, and assistance than did white communities.

It is not too difficult to understand why low income communities experience higher social vulnerability—they have limited or no ability to protect themselves from risk and to bounce back after a disaster.[20] Alice Fothergill and other sociologists explain: "Recovery is more difficult when the household has a low income to begin with, their employment may tend to be more disrupted by a disaster, the household may have little or no savings, and may not have adequate or any insurance."[21]

Professor Cutter adds that minority populations, like low-income populations, are also more vulnerable to personal injuries and property damage when they are exposed to the same magnitude event as non-minority populations.[22] During and after Hurricane Katrina, "the city's [black] poor were left behind to die . . . or to survive for days under conditions so appalling . . . that some observers called [it] 'ethnic cleansing by inaction.'"[23]

But the reasons why minority communities experience higher vulnerabilities are somewhat opaque. People like the Sioux tribes in the Midwest and the residents of the Lower Ninth Ward and St. Bernard Parish are most vulnerable to the vagaries of government disaster preparedness and response policies because they have minimal access to political representation, information, technology, and social networks. Fothergill agrees that when disasters occur in the United States, racial and ethnic communities suffer more due to language barriers, housing patterns, community isolation, and cultural insensitivity. She explains:

> Residential segregation patterns and insurance company red-lining practices may also contribute to minority households having less than optimal insurance, or for being insured with non-major insurance firms. Racial and ethnic minority households also tend to have less access to information about relief assistance and opportunities. More affluent and non-Hispanic white households are more likely to know how to "work the system," to fill out forms, ferret out information, and navigate through convoluted government procedures than the minority households, resulting in the minority and lower-income households receiving much less relief aid, and recovering economically much more slowly.[24]

Why do certain populations choose to live in high-risk areas in the first place? Of course, the socio-demographic characteristics of populations residing in flood-prone areas often reflect historical settlement patterns, which sometimes have little to do with race, ethnicity, or even poverty. Within any given region, there may be no clear relationship between socio-economic status and elevation. In some places, wealthy landowners reside on the coast or along the river banks and enjoy their expansive views of the water, while in others, the lands along the waterways are populated by communities made up of African American, Asian, Hispanic, Native American, or other minorities. With respect to the latter groups, choice may have little to do with it—flood-prone land is often cheaper.[25]

The problems associated with social vulnerability are exacerbated when the business leaders and decision-makers don't experience the same effects as the less fortunate members of the population. In his book *Collapse: How Societies Choose to Fail or Succeed,* Jared Diamond, recipient of the National Medal of Science for his discoveries in evolutionary biology and history, explains why. A common theme throughout the book is how the insulation of the decision-makers from the consequences of their actions—in other words, the imposition of the ecological and economic burdens of their actions on the political "have-nots"—has ultimately resulted in societal collapse. According to Diamond, "In societies where the elites do not suffer from the consequences of their decisions, but can insulate themselves, the elite are more likely to pursue their short-term interests, even though that may be bad for the long-term interests of the society."

The Maya civilization, which reached improbable heights, is a leading example. The Maya ruling class demanded that the forests, the rivers, the fish, and other natural resources be harvested and ultimately exhausted to ensure the rulers' own supremacy.[26] Diamond continues, "Maya kings were consumed by immediate concerns for their prestige (requiring more and bigger temples) and their success in the next war (requiring more followers), rather than for the happiness of commoners or of the next generation."[27] After centuries of prosperity, their civilization collapsed, and their settlements were reclaimed by the lush jungles of Central America and southern Mexico.

Conversely, where decision-makers themselves must live with the consequences of their actions, they tend to make more sustainable, equitable decisions. Diamond points to Holland as an example:

In Holland everybody lives in the Polders [lands below sea level that have been drained, reclaimed from the sea, and surrounded by dykes], whether you're rich or poor. It's not the case that the rich people are living high up on the dykes and the poor people are living down in the Polders. So when the dyke is breached or there's a flood, rich and poor people die alike. . . . [R]ich people cannot insulate themselves from consequences of their actions. They're living in the Polders and therefore there is not the clash between their short-term interests and the long-term interests of everybody else.[28]

Holland may be atypical. In the United States, the evidence suggests that minorities and the poor do in fact suffer disproportionately from floods.

Are We There Yet?

IN 1992, AROUND the same time the Yankton Sioux were told that their ancestors had begun floating away from the White Swan Cemetery, the U.S. Environmental Protection Agency issued its report on *Environmental Equity: Reducing Risk for All Communities*. Instead of running from the earlier United Church of Christ report, the EPA openly acknowledged that minority and impoverished populations shoulder greater environmental burdens and health risks than others:[29]

Historically, people of color communities have borne a disproportionate burden of pollution from landfills, garbage dumps, incinerators, smelters, sewage treatment plants, chemical industries, and a host of other polluting facilities. Many dirty industries have followed the "path of least resistance," allowing low-income and people of color neighborhoods to become the "dumping grounds" for all kinds of health-threatening operations.[30]

The promotion of environmental justice—or at least the public discussion of environmental *injustice*—became a fundamental part of federal decision-making processes when President Clinton issued Executive Order 12898 in 1994. This order directs federal agencies to consider whether poor or minority communities would suffer inordinately from federal actions involving the environment.[31] More specifically, the executive order requires agencies to collect data on low-income and minority populations that may be disproportionately at risk. It also calls for improved methodologies for assessing and mitigating impacts to various groups, including Native Americans and other subsistence hunters and fishers. Most importantly, it seeks the "fair treatment and meaningful involvement of all people regardless of race, color, national origin, culture, education, or income with respect to the development, implementation, and enforcement of environmental laws, regulations, and policies." Under the order, *fair treatment* means that "no group of people, including racial, ethnic, or socioeconomic groups, should bear a disproportionate share of the negative environmental consequences resulting from industrial, municipal, and commercial operations or the execution of federal, state, local, and tribal environmental programs and policies."[32]

With all this attention to environmental justice, surely things have improved since Warren County made national headlines, and since the Missouri River dams displaced so many Native American people and their ancestors' graves. Or maybe not. As Dr. Bullard explains:

> [E]nvironmental injustice in communities of color is as much or more prevalent today than two decades ago. Racial and socioeconomic disparities in the location of hazardous waste facilities are geographically widespread throughout the country. . . . [I]n 2007 communities of color were more concentrated in areas with commercial hazardous sites than in 1987. Even when statistical analyses take socioeconomic and other non-racial factors into account, race continues to be

a significant independent predictor of commercial hazardous waste facility locations.[33]

The plight of Esther Williams and other residents of her hard-hit Ninth Ward neighborhood, detailed above in chapter 8, vividly illustrates the environmental injustices perpetuated by all levels of government just a few years ago, during and after Hurricane Katrina.

10

DOUBLE-TAKES

CHARGING TAXPAYERS, TWICE

I ronically, the nation's flood strategies shift the risk *away* from those who gamble on floodplain development or otherwise settle in vulnerable areas. As a result, many of the risk-takers avoid paying the full price of their behavior; instead, they shift the costs onto others. As described in the previous chapter, too often those who bear the brunt of the nation's failed flood management policies are the poorest members of society, who lack the financial means to settle in safer areas, to evacuate from a floodplain during disaster, or to purchase flood insurance.

But apart from these risk-takers, both voluntary and involuntary, it is the taxpayers who subsidize the entire enterprise. And not just once, but twice. First, tax dollars finance the massive federal machine that encourages risky settlement: levees, federal flood insurance, and disaster relief. And second, in those cases where laws discourage risk-taking through land-use regulations, building prohibitions, wetlands protection, and the like, some landowners demand that taxpayers compensate them for complying with the law. This odd turn of events —paying some people to follow the law—is the result of a controversial and complicated legal theory known as the *regulatory takings* doctrine. That doctrine is the offspring—some would say, the illegitimate child —of its better known parent, the power of eminent domain.

Damned If You Do: The Cost of
Forbidding Risky Behavior

WHEN CONGRESS ADDED floodways to the nation's arsenal of flood control weapons, it knew that farms and rural communities had already been established in areas that the Corps would likely select as emergency overflow areas. Invoking the federal government's power of eminent domain, Congress ordered the Corps to condemn those properties to acquire title, or at least flowage easements granting the right to periodically flood the property.

The condemnation process can be contentious and difficult. For example, as chapter 4 described, it took the Corps decades to acquire flowage easements and other property rights in the area that would become the Birds Point-New Madrid Floodway below Cairo, Illinois.

In essence, eminent domain is a forced sale from an unwilling seller to the government, designed to facilitate public uses such as spillways, dam construction, road building, railroad development, power lines, and other public projects that require the acquisition of title to large tracts of land (or easements for the use of private property, as in the case of floodways). It is clear that the drafters of the Fifth Amendment to the U.S. Constitution intended for the government to pay compensation when it physically appropriates private property through the exercise of eminent domain and takes title to the property or to an easement.

Eminent domain spawned the controversial "regulatory takings" doctrine. Cited sporadically in the early twentieth century, the doctrine gained momentum in the 1980s. By that time, the nation had learned the value of natural wetlands and floodplains, perhaps the most effective means available to tame floodwaters. It had also learned the value of laws designed to protect wetlands and floodplains, and to prevent unwise development in risky places. But the public interest in flood protection would soon collide with the private interests of individual

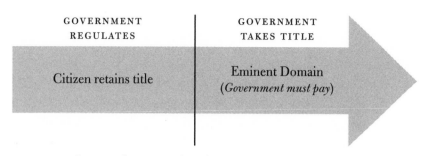

GOVERNMENT
REGULATES

GOVERNMENT
TAKES TITLE

Citizen retains title

Eminent Domain
(*Government must pay*)

Conceptual representation of the doctrine of eminent domain
Diagram by Christine A. Klein, 2012

and corporate landowners. Both interests are critical, but it's tough to advance one without affecting the other.

Using the regulatory takings doctrine, the U.S. Supreme Court began to side with the landowners. The legal movement was fueled by a growing social movement of the 1980s that expressed antipathy toward the government (especially the federal government) and that elevated individualism and the private sector. Tapping into this trend, the Supreme Court began to require the government to pay landowners when their property was merely *regulated,* but title (or an easement) was not actually taken over by the government. To support this evolving concept, the Court reasoned, some government regulations are so onerous that they are the practical equivalent of eminent domain. If the government must pay when it takes over private property (title and all), the Court concluded, then the government should also pay when it regulates heavily, even if landowners retain title to their property.

In 1922, U.S. Supreme Court Justice Oliver Wendell Holmes issued an influential opinion on this topic in the case of *Pennsylvania Coal Co. v. Mahon.* In that dispute, the Mahon family owned their home and the surface of their lot, but the Pennsylvania Coal Company owned the coal beneath it and the right to mine that coal deposit. Pennsylvania passed a law, the Kohler Act, that forbade the mining of coal in such a way as to cause the subsidence (or sinking in) of the land surface in places where coal seams supported, among other things, buildings used for human habitation. The Mahons sued the coal company to prevent it from mining beneath their property in a way that would cause their land to cave in and their house to collapse. In defense, the coal company claimed that the Kohler Act was unconstitutional because it deprived the company of its subsurface coal without paying for it. The Court sided with the coal company. Because the Mahons had "seen fit to take the risk of acquiring only surface rights," Justice Holmes wrote, "we cannot see that the fact that their risk has become a danger warrants the giving to them greater rights than they bought." In an often-quoted, but notoriously unhelpful statement, Justice Holmes established the general rule "that while property may be regulated to a certain extent, if regulation goes *too far* it will be recognized as a taking" (italics added).

Of nine justices, only Justice Louis D. Brandeis disagreed with Justice Holmes. Explaining the basis for his dissent, Justice Brandeis argued, "Coal in place is land; and the right of the owner to use his land is not absolute. . . . If by mining anthracite coal the owner would necessarily unloose poisonous gasses, I suppose no one would doubt the power of the State to prevent the mining, without buying his coal fields." Brandeis asked, "And why may not the State, likewise, without paying compensation, prohibit one from digging so deep or excavating so near the surface, as to expose the community to like dangers? In the latter case, as in the former, carrying on the business would be

GOVERNMENT REGULATES		GOVERNMENT TAKES TITLE
Not "too far" (*No compensation required*)	"Too far" (*Government must pay*)	Eminent Domain (*Government must pay*)

Conceptual representation of the regulatory takings doctrine
Diagram by Christine A. Klein, 2012

a public nuisance." Although the Supreme Court would later accept Justice Brandeis' views and incorporate them into a well-recognized "nuisance defense" to takings claims, Justice Brandeis was unable to convince a majority of his brethren that the Mahons should be protected from harmful activities by the coal company.

Where should courts draw the line between government regulations that require compensation, like the Kohler Act, and those that do not? The better part of a century has passed since the court's decision in *Pennsylvania Coal,* but the regulatory takings debate rages on. Numerous judges and scholars have struggled to define just when a regulation goes "too far" and when it stops short of crossing that ill-defined line. Others believe that the entire regulatory takings doctrine is an illegitimate judicial creation, and that the government should not be required to pay for regulations that do not confiscate title to private property. Legal writers have variously described the regulatory takings doctrine and its defenses as a "muddle,"[1] a "puzzle,"[2] and a "knot."[3] The U.S. Supreme Court itself made an embarrassing analytical detour for a period of decades, during which it sometimes intermingled the regulatory takings doctrine with an essentially unrelated legal theory known as "substantive due process" analysis. In unanimously reversing course in 2007, the Court acknowledged its mistake and conceded that it must "eat crow" to correct its error.[4]

To explain the confusion, it is helpful to think about the extraordinarily difficult task with which the regulatory takings doctrine has been saddled: balancing private rights against the public interest. Viewed from that perspective, it is no wonder that judges have trouble deciding in any particular case how to accommodate both individual property rights and the community's interest in protecting the public health, safety, and welfare. As Justice Holmes admitted back in 1922 (even though he ruled against the government in *Pennsylvania Coal*), "Government hardly could go on if to some extent values incident to property could not be diminished without paying for every such change in the general law. As long recognized, some values are enjoyed under an implied limitation and must yield to the police power." But, on the other hand, the Supreme Court has interpreted the Fifth Amendment as a constitutional requirement "designed to bar Government from forcing some people alone to bear public burdens which, in all fairness and justice, should be borne by the public as a whole."[5]

Not surprisingly, some of the most difficult legal line-drawing challenges have arisen in the context of physical settings where nature draws the sometimes blurry line between land and water. Floodplains present one such ecological transition zone, as illustrated by the Court's opinion in *First English,* discussed in chapter 6, which considered the constitutionality of Los Angeles County's flood control ordinance that prevented a church from rebuilding its church camp after a flood. Likewise, coastal areas and wetlands have been the subject of prominent regulatory takings disputes, as demonstrated by the Court's opinion in *Lucas v. S. Carolina Coastal Council,* which involved a challenge to a South Carolina zoning ordinance that imposed construction setbacks from sensitive dune areas, in part to minimize hurricane damage.

Like *First English,* the *Lucas* lawsuit illustrates the phenomenon of *double-takes*. Because South Carolina lost the lawsuit, its taxpayers were responsible for paying Lucas to abide by the law, and to refrain from building on his two lots. But that would not be the first time that taxpayers (federal, state, and local) compensated Lucas. They had been paying for years to subsidize development on the shifting sands of the Isle of Palms. In fact, construction of the Wild Dunes Resort likely would not have been possible but for taxes that paid for a bridge to the mainland, as well as repeated beach nourishment measures and sandbagging. In addition, taxes subsidize the federal flood insurance program and disaster relief that serve as a backstop for development. Given the history of subsidies that benefitted Lucas and his resort, one can't help but think that the taxpayers should not be charged yet again, this time to enforce the law. It is no wonder that the Supreme Court was unable to deliver a unanimous opinion—six justices sided with Lucas, and three justices agreed with the taxpayers.

Damned If You Don't: The Cost of
Allowing Risky Behavior

ALTHOUGH THE REGULATORY takings doctrine can make it costly for taxpayers when laws forbid risky behavior, it is also expensive to allow —and even facilitate—development in vulnerable areas. Since 1928, the country has spent hundreds of billions of dollars on flood control structures in coastal and floodplain areas, on flood insurance subsidies, and on disaster relief. Despite these massive expenditures, economic losses due to flooding have more than doubled during the same time period, currently costing about $6 billion annually. According to testimony before the Senate Committee on Environmental and Public Works in 2005, "Because so many Corps flood control projects induce development in harm's way, flood damages have more than tripled in real dollars in the past eighty years—even as the Corps has spent more than $120 billion on flood control projects."[6]

The mighty Mississippi River presented a challenge that no self-respecting engineer could resist. The nation's Army Corps of Engineers stepped in, engaging in a type of domestic war to control the river. In the era before the development of the ecological sciences and conservation biology, there was a widespread failure to appreciate the broad, interconnected nature of rivers and their floodplains, or oceans and their adjacent beach, dune, and barrier island complexes. Rather, the compartmentalized thinking of the time was unable to recognize the relationship between flood control, navigation, and development; floods were viewed as isolated acts of God, without consideration of the human responsibility for magnifying flood damage. As a result, society remained uneasy about the federal government's constitutional authority to control floods, and was therefore content with the Army Corps of Engineers' early, constrained "levees only" role. This approach proved to be disastrous, as taxpayer-funded levees strait-jacketed the river and shifted the risk of flooding to other riverfront communities not yet protected by levees (or behind levees that failed).

The flood of 1927 brought international attention to the lower Mississippi. The nation began to acknowledge that the disaster had been magnified greatly by the existence and failure of levees, transforming natural flooding into a "manmade disaster." But instead of rejecting engineered flood control, the nation ultimately called for an even more elaborate system of structures that would include floodways and retention areas expected to provide a safety valve for the overflow of levee-constricted rivers. The bulk of the responsibility for flood control was then placed on the federal government, which in turn delegated responsibility to the Army Corps of Engineers. The Corps proceeded zealously, virtually unfettered by legal limits that would not appear until the enactment of the Administrative Procedure Act of 1946 (which, as its name suggested, established procedures and limits to control federal agencies) and environmental laws later in the twentieth century. Thus, floodplain communities shifted the risk of vulnerable development to federal taxpayers and supported the construction of engineered structures that often proved to do as much harm as good.

As a mid-century sequel to the flood of 1927, flood waters claimed towns and fields in the Midwest in the 1940s and then again in the 1950s. Congress authorized yet more structural flood control and created two additional subsidies for floodplain development through the Disaster Relief Act of 1950 and the National Flood Insurance Act of 1968. Laws such as these shifted the risk away from floodplain inhabitants and toward taxpayers.

Risk-shifting subsidies continued with renewed force following the 1993 flood. By late century, there was little local appetite for controlling floodplain and coastal development. This became apparent in the post-flood analysis, which revealed numerous loopholes in the flood insurance program that weakened compliance with federal requirements for local land-use regulations. Moreover, the post-war building boom had morphed into the phenomenon of suburban sprawl, which prompted expansion into a variety of undeveloped areas, including floodplains. On top of all this, the "property rights" movement began in earnest about 1985, building on the regulatory takings doctrine recognized by the U.S. Supreme Court in *Pennsylvania Coal, First English, Lucas,* and other cases. As a result, land-use regulators met with a powerful deterrent—a determined group of advocates who used the Fifth Amendment as a constitutional shield against government regulation. Consequently, floodplain construction (and reconstruction) continued largely unabated. In St. Louis, for example, the nation's largest strip mall sprouted up on the very site that had been catastrophically inundated by floodwaters in 1993. Thus, developers continued to enjoy the benefits of sprawling into vacant land, while shifting the risk onto taxpayers for flood control structures, insurance, and disaster relief, and also discouraging local regulation with the threat of takings litigation.

Hurricane Katrina illustrated many of the same lessons. But these lessons also introduced a new dimension to this risk-shifting history, pitting the infrastructure-intensive interests of the shipping industry and the wetlands-destroying interests of the oil and gas industry against the interests of urban dwellers, many of them poor and black. Often, hurricane victims might not have willingly chosen to live in the floodplain or to forgo federal flood insurance, but they might not have had any other realistic financial option. In those cases, the risk of failed structures was borne by a segment of the population less able to do so (at least in a financial sense) and less likely to have the means to evacuate in advance of the storms.

Accepting Responsibility, Acknowledging Risk

AFTER THE MIDWESTERN flood of 1993, the Galloway Report issued a strong call for risk-takers to accept responsibility for the costs and consequences of their actions. At the very least, the report suggested the elimination of subsidies that skew the operation of the free market by subsidizing, insuring, protecting, and compensating those who choose to settle in notoriously risky areas (such as places within flood zones mapped by the Federal Emergency Management Agency). A similar call for a fair reckoning underlies periodic efforts to reform the regulatory takings doctrine. Reformers acknowledge that the government can over-reach, but also recognize the importance of regulations that protect the public health, safety, and welfare. Likewise, landowners must feel secure in their reasonable expectations of enjoying their property rights, but they should not be allowed to use their land in a way that unfairly shifts risk (physical or financial) to others.

Contrary to *Lucas'* view that undeveloped floodplain and coastal land is "valueless" in its natural state, some judges have begun to recognize the important functions performed by floodplains, coastal dunes, and wetlands. Such "ecosystem services" are immensely valuable, and include controlling floods, filtering out water pollution, providing habitat for fish and wildlife, stabilizing the climate by storing carbon, and providing aesthetic, recreational, and educational opportunities. Because these services are so important, some courts view as "nuisances" development activities that destroy the natural functions of floodplains, coastal zones, or wetlands. Under this rationale, laws that protect aquatic resources do not create regulatory takings. Instead, it is the destruction of such resources through risky development that "takes" from the rest of society, in terms of foregone ecosystem services and taxpayer subsidies. This is the same notion suggested by Justice Brandeis in his *Pennsylvania Coal* dissent, when he recognized a "nuisance defense" to regulatory takings lawsuits.

Three cases—all decided in coastal states prone to hurricanes—are representative of this emerging line of analysis. First, the Massachusetts Supreme Court has recognized significant value in natural coastal floodplains, even when subject to a total development ban. Thus, in the 2005 dispute *Gove v. Zoning Board of Appeals of Chatham*, the court held that a state regulation banning without exception all new residential construction within the 100-year floodplain did not require compensation as a regulatory taking. Finding that the regulation did not deprive the property of all economic value, the court noted that the property was still valuable for fishing, shellfishing, recreation, utility installation, and agriculture. The court dismissed the landowner's residential development aspirations as unreasonable when measured against the property's status as "highly marginal, . . . exposed to the ravages of nature, [and] . . . for good reason [left] undeveloped for several decades even as more habitable properties in the vicinity were put to various productive uses."[7]

In neighboring Rhode Island, the court recognized floodplain and coastal construction as a nuisance that can be prohibited by law without compensation. In its 2005 decision in *Palazzolo v. Rhode Island,* the Rhode Island Superior Court upheld the denial of a permit to fill in eighteen acres of coastal salt marsh as part of a proposed beach-front development. Rejecting a regulatory taking claim, the court found that the law merely prevented the landowner from creating a public nuisance by destroying wetlands. As such, it was the landowner who proposed to "take" something of value from his neighbors, and not the law that would have taken away rights to which the landowner was entitled.

Finally, in 1988, *Adolph v. FEMA* resolved a dispute concerning an area that would be devastated in 2005 by Hurricane Katrina. A group of Louisiana property owners challenged strict flood control regulations passed by the Plaquemines Parish Commission Council, alleging an unconstitutional regulatory taking. In upholding the challenged regulations, which had been adopted in order to participate in the National Flood Insurance Program, the court issued a strong endorsement of the NFIP. The court was favorably influenced by the nature of floodplain management, which it viewed as an effort to prevent some landowners from imposing nuisance-like dangers on others. As the court concluded, "Flood-hazard zoning and other regulations serve a vital purpose in protecting the people who occupy the regulated land and in protecting neighboring landowners from increased flood damage and in protecting the general public." Such regulations, the court noted, involve "the safety of lives and property, not merely environmental or aesthetic considerations."[8]

*   *   *

THE HISTORY OF federal flood control, together with the history of the regulatory takings doctrine, paints a bleak picture for taxpayers and floodplain occupants alike. The concluding chapter looks for glimmers of hope, highlighting promising new approaches being tested by federal, state, and local governments.

CONCLUSION

HOW LAW HAS HURT,
HOW LAW CAN HELP

They huddled nervously beside the river.

"Will we get wet?"

She couldn't help but smile as she responded, "Yes."

"Will we end up where we started?"

She paused for a moment. The answer seemed obvious enough: "No."

She was working as a summer raft guide in Colorado, taking a break after her first year of law school. Her customers were understandably anxious that June morning as they stood on the banks of the Arkansas River and snugged up their life jackets. The river raged past them, swollen with runoff from barely melted mountain snow as it made the fourteen-hundred-mile journey down to where it joined the Mississippi River.

The questions were always the same. Worried customers found it difficult to remember that rivers flow downhill. *No,* the group would not end up where it had started, although the company van would be waiting at the river take-out to drive the customers back to their cars. And *yes,* they were guaranteed to get wet as they navigated the swirling rapids.

In a sense the questions were flattering, implying that she had the superhuman ability to power the fourteen-foot raft and seven adults upstream against the current, or that she could finesse every breaking wave of the technically challenging Class IV run so that the group would stay dry. But from another perspective, the questions were troubling. Only in Disney World, she thought, was the environment so controlled that wave height could be calculated and rivers could indeed flow in circles to end where they began. And most people today, she mused, were probably more familiar with water parks and Disney World than with natural rivers. Author Richard Louv got it right when he described our children's disconnection from the natural world as "nature deficit disorder." In *Last Child in the Woods: Saving Our Children from Nature-Deficit Disorder,* Louv suggested that schools should incorporate outdoor nature education into the curriculum. He wrote, "Passion is lifted from the earth itself by the muddy hands of the young; it travels along grass-stained sleeves to the heart."[1]

But is the Arkansas River—and the Mississippi itself—really so natural? After all, both are constrained by numerous levees and dams that affect everything from the speed of the current, to the volume of sediment flow, to the temperature of the water.

But these big questions would have to wait. This particular group, at least, had ventured out into nature, and they were approaching Maytag Rapid. It was time to get the raft into position.

* * *

MANY YEARS LATER, the Arkansas River rafting guide wrote this book in collaboration with a like-minded colleague. She still ponders the questions asked by the customers of the rafting company: "Will we get wet?" "Will we end up where we started?" Applied broadly, those questions serve as a metaphor for the past century's management (and mismanagement) of the Mississippi River basin. The answers depend, as this book suggests, on whether or not we learn from our past mistakes.

It comes down to this: Actions have consequences. If we acknowledge that some tragedies are of our own making—that they are *unnatural disasters,* not uncontrollable "acts of God"—then we have a fighting chance at making better decisions in the future. It's not about blame, although in some cases the courts can reinforce important lessons by holding accountable those who exacerbate natural disasters through negligence, mistake, incompetence, or the like, and by compensating the victims of those misdeeds.

Looking Back: How Law Has Hurt

THE PROBLEM IS not that we fail to learn from our mistakes. Sometimes we do learn a thing or two, and sometimes we even take tentative steps to avoid the same mistakes in the future. Instead, the problem is that we have short attention spans, lapse into complacency, or run off to address the next pressing issue in the calm period between the storms.

We have learned much during our experiment with intensive federal flood management, which began in earnest with the passage of the Federal Flood Control Act of 1928. Three lessons stand out: 1) Rivers *will* flood; 2) levees *will* fail; and 3) unwise floodplain development *will* happen if we let it.

* * *

FIRST, RIVERS *will* flood.

The Mississippi continues to rise above its banks, despite the corset of levees designed to prevent it from spilling into its floodplain. As the U.S. Geological Survey reported, during the twentieth century floods topped the list of "natural disasters" in terms of death and property damage.[2] We pour billions of dollars into control structures: The Army Corps of Engineers has spent in excess of $120 billion on flood projects since 1925. But despite our best efforts, flood damages continue to rise.[3]

As the Association of State Floodplain Managers warns, "Building in a floodplain is like pitching your tent on a highway when there are no cars coming." We flirted with—in fact, we married ourselves to—the disastrous "levees-only" policy. But we learned from the flood of 1927 and rejected that policy, in the words of one official, as "the most colossal blunder in engineering history." To replace the natural floodplains that had been walled off behind levees, the Corps designed artificial "floodways." In a poor imitation of natural processes, we now direct overflow to these engineered floodways through gaps (*fuse plugs*) where the levee height is lower than in surrounding areas. In a pinch, we use dynamite to blast out an emergency exit for the river. In effect, we're still putting our faith in levees, but we've added a few more tools to the box.

We've also learned that flooding is not such a bad thing after all, eco-logically speaking. It enriches floodplain soils and delivers sediments to build up river deltas. Just as we devised engineered substitutes for natural floodplains, we also use technology to mimic natural floods and to produce at least some of their benefits. To restore levee-starved riv-erine habitats, at times we deliberately manipulate dams and reservoirs to create "managed" floods.

Levees, floodways, spillways, fuse plugs, dynamite, managed floods. In many cases, wouldn't it have been easier—and smarter—just to let the river reclaim its floodplain and to settle out of harm's way?

*   *   *

SECOND, LEVEES *will* fail.

As one engineer explains, there are two kinds of levees: "Those that have failed and those that will fail!"[4] In fact, levees are *designed* to fail under some circumstances. In engineer-speak, this is dubbed *residual risk*. The typical "100-year levee" is engineered to hold back the level of flooding that has at least a 1 percent chance of occurring each year (which translates into a 26 percent chance of flooding over the life of a thirty-year mortgage). Beyond that, all bets are off. The American Society of Civil Engineers (ASCE) warns, as an "essential levee fact," that no levee is flood-proof. In the ASCE's words,

> Levees *reduce* the risk of flooding. But no levee system can *eliminate* all flood risk. A levee is generally designed to control a certain amount of floodwater. If a larger flood occurs, floodwaters will flow over the levee.

Flooding also can damage levees, allowing floodwaters to flow through an opening, or breach.[5]

* * *

THIRD, UNWISE FLOODPLAIN development *will* happen if we let it.

And we let it happen by failing to adopt appropriate land use requirements and by providing subsidized flood insurance.

Congress created the National Flood Insurance Program (NFIP) in 1968, but it did so with trepidation. Our legislators worried that rather than limiting losses, the availability of subsidized insurance policies would cause further development in the floodplains and lead to even greater flood damage.[6] (After all, the private market refused to provide flood insurance for good reason—such policies were far too risky.) Just two years earlier, a national task force had warned, "A flood insurance program is a tool that should be used *expertly or not at all*. Correctly applied, it could promote wise use of flood plains. Incorrectly applied, it could exacerbate the whole problem of flood losses." The task force's warning proved apt. Attempting to wield the NFIP tool "expertly," Congress adopted significant amendments to the program's legislation in 1973,[7] 1994,[8] and 2004.[9]

The 1993 floods and the 2005 hurricane season illustrated critical flaws of the NFIP: People had in fact been allowed to build their homes and businesses in floodplains; not enough of them bought into the insurance pool (even though they may have been legally required to do so); and insurance premiums were not high enough overall to make the program financially sustainable in the long term.[10] For the three and a half decades leading up to Hurricane Katrina, the NFIP was able to support itself with premiums and fees generated by insurance policies, despite satisfying claims for flood after flood, including paying out almost $300 million after the Midwest flood of 1993. But after Hurricane Katrina, over $16 billion worth of claims swamped the NFIP's ability to remain financially self-supporting.[11] To make up the shortfall, the NFIP obtained an interest-accruing loan from the

U.S. Treasury. By January 2011, the program owed the treasury almost $18 billion.[12]

Congress designed the NFIP to steer development away from high-risk areas. But over time, it became apparent that this goal had not been realized. The case of so-called *repetitive loss* properties illustrates this phenomenon. During the NFIP's first four decades, 50,644 insured properties sustained flood damage on more than one occasion. Of these, 11,706 had sustained four or more losses, or had sustained two or more losses, the cumulative payments for which exceeded the value of the property. Rather than move out of the floodplain or take measures to flood-proof their properties (such as elevating them), these landowners collected insurance benefits and stayed put. Although repetitive loss properties held only 1 percent of all NFIP policies, they sucked up an annual average of 30 percent of all claims payments. From 1978 to 2004, these determined floodplain dwellers pocketed at least $2.7 billion in benefits, with estimates reaching as high as $4.6 billion.[13]

These repetitive loss properties are damning evidence that the NFIP has failed to reduce flood damage or to dissuade people from occupying hazardous areas. Instead, it has enabled a small (but expensive) group of people to remain stubbornly in harm's way.

Congress attempted to address some of the NFIP's shortfalls when it passed reauthorization legislation known as the Flood Insurance Reform Act of 2012. The law was aimed at making the NFIP fiscally stable and sustainable by, among other things, gradually phasing out below-market subsidized insurance rates, and raising the cap on permissible annual rate increases. No sooner than the legislation passed, however, Hurricane Sandy ravaged portions of the Atlantic coastline, destroying or damaging some 385,000 homes in New Jersey and New York. When it became apparent that the strict new law would have a severe financial impact on already-devastated homeowners—in some cases requiring annual insurance premiums of up to $20,000—Congress blinked. Just one year after Congress passed the reform legislation, it began debating potential measures to delay or preclude implementation of the new provisions that had encountered opposition.

*   *   *

THE FEDERAL GOVERNMENT has promised more than it can deliver —that it will somehow hold back the mighty Mississippi River, keeping all citizens safe and dry, no matter where they have chosen to put down stakes. The mindset is self-reinforcing. Accustomed to federal assurances, we now insist on a sanitized and risk-free world. Floodplains have become an inconvenience; getting wet is unheard of. Modern rivers are mostly unnatural human creations, and when they reassert their natural impulses—as they inevitably do—the nation suffers.

The currents of our flood control policy have swirled back around, coming to pool at a fitting place—the National Mall in Washington, D.C. The U.S. Supreme Court and Congress sit at the far eastern end of the Mall, on opposite sides of First Street. Here, the Court serves as the final arbiter of many high-level disputes, including flood liability and Fifth Amendment takings claims, and Congress enacts laws on all topics, including those that affect river and flood management. The U.S. Army Corps of Engineers operates its national headquarters nearby, at 441 G Street. The White House is a dozen or so blocks to the west. The president, commander in chief of the United States, is also the commander of the Army Corps of Engineers.

Surely, these pillars of national flood policy are constructed firmly on solid—maybe even sacred—ground, aren't they?

As it turns out, portions of Washington, D.C., were built on a swamp, although some insist that "tidal pool" is a more accurate description of the sometimes soggy terrain. In any case, since the late 1930s the Potomac Park levee has stood guard between the Lincoln Monument and the Washington Monument, ready to hold back the Potomac River from the city. Its contours blend in well enough with the manicured landscape of the National Mall. But it had two dangerous gaps—at

Twenty-Third Street NW and at Seventeenth Street NW. When flood-waters threaten, workers plug these gaps with sandbags. They could also fortify the area with a temporary earthen berm, using soil bulldozed from the grounds of the Washington monument.

The Corps of Engineers and the National Park Service (which manages the National Mall and whose headquarters, incidentally, are located in Foggy Bottom, so-named, some say, for its low-lying, fog-producing, swampy ground) aren't taking any chances. After Hurricane Katrina gave the nation a blaring wake-up call, FEMA contemplated including a large area of downtown Washington in its updated 100-year floodplain maps. According to FEMA's calculations, during a severe flood, up to ten feet of water could cover parts of downtown and the Mall, including critical institutions and treasured landmarks. If the maps were updated as planned, then occupants of the revised floodplain zone would be pressed to buy federal insurance if they wanted to qualify for federally backed mortgages or to receive insurance benefits in the wake of a flood. To forestall these consequences, officials decided to beef up the area's flood protection by filling in the Seventeenth Street gap of the existing Potomac Park levee.

Design options ranged from the mundane to the elaborate. Planners wanted to choose an approach that would blend well into the National Mall, an area the Corps described as "America's front yard." After several years of planning, the National Capital Planning Commission selected a two-part scheme. First, the plans called for eight-foot high concrete floodwalls on either side of the Seventeenth Street levee gap. Second, the plans included a removable metal wall about 140 feet in length and about 9 feet tall. According to the plan, when floodwaters threaten, the National Park Service should take the "pop-up floodwall" from storage in its newly constructed underground bunker and quickly set it up between the Seventeenth Street floodwalls. By design, the new levee will protect the area only from *river* flooding. As FEMA admits, it is not intended to protect from other flood threats, such as a tidal storm surge moving in from the Potomac (like the one that accompanied

Construction to close the gap in the Potomac Park levee at Seventeenth Street
Photograph by Christine A. Klein, 2012

Hurricane Isabel in 2003) or precipitation heavy enough to overcome the city's aging sewer system.

Overall, the Potomac Park levee—in its original or fortified form—is a metaphor for our national flood policy: It resembles a band-aid more than a life-saving procedure. Superficially, it looks impressive. Let's just hope that emergency workers can act quickly enough when it is time to fire up the bulldozers or to pop the 140-foot temporary floodwall into place.

* * *

ACROSS THE POTOMAC and not far downstream, the merchants and bartenders of Old Town Alexandria view flooding a little differently. When it happens—as it does fairly regularly, according to Richard Baier, director of Alexandria's Department of Transportation and Environmental Services—it's just "typical nuisance flooding. . . . Usually once or twice a month we have minor flooding at the foot of King Street." If need be, local residents and business owners can pick up a few sand bags from the city truck parked at the corner of King and Lee streets. Long-time residents view this as "normal operating procedure." It can be somewhat vexing when the tide comes in during peak business hours, making water levels rise even farther, but people find ways to cope. When the Potomac rose two to three feet higher than usual during the wet spring of 2010, Mark Kirwan, the owner of O'Connell's Irish pub, said, "We're used to it. The bar is open, people need to drink." Locals and tourists alike find themselves wading shin-deep to get to O'Connell's and other popular taverns. For the non-drinkers, Ben and Jerry's Ice Cream is just down the street, across from the Old Dominion Boat Club (where you can savor your Chubby Hubby double-dip cone while perusing the boats in the marina). The shops stay open late, in case you need a new pair of shoes to replace the sodden ones you've been wearing, or if you want to pick up a souvenir or two for your friends back home. Rachael Bright, who has worked at the Silver Parrot Boutique on lower King Street for thirteen years, says that she's seen lots of floods. "I think it's actually a little bit of a draw. [People] get curious. They come out and snap pictures. And maybe afterward they go shopping or go eat."[14]

Moving Forward: How Law Can Help

LEGAL REFORMS ARE needed to hold accountable those who endanger themselves or others, both physically and financially. Unlike some of the measures adopted by the Corps, these reforms are institutional, not structural. We propose the following: (1) Eliminate federal immunity for flood damage; (2) reform the decision-making metrics for the construction of flood control projects; (3) reform the Fifth Amendment takings doctrine; and (4) reform the National Flood Insurance Program both to keep people out of harm's way and to force those who engage in risky behavior to assume the risk (and the cost).

(1) *Eliminate federal immunity for negligent acts that cause or exacerbate flood damage.* The 1928 Flood Control Act immunizes, or protects, the federal government from any liability "of any kind . . . for any damage from or by floods or flood waters at any place." Although Congress was willing to authorize widespread federal involvement in flood control, a responsibility that it had previously considered a purely local affair, Congress was only willing to go so far: If well-intended federal action caused or exacerbated the damages, well, that was just too bad for those unfortunate people who got hurt.

Just one year after Hurricane Katrina struck, the Corps admitted culpability for the design and construction flaws that led to the devastation of New Orleans. But the Corps also claimed that sovereign immunity shielded it from claims brought by individuals and businesses whose properties were destroyed when the levees failed. In the Katrina litigation described in chapter 8, federal district Judge Stanwood Duval rejected the Corps' immunity argument. While the appeals court agreed with him at first, it changed its mind just six months later. This unsettled response from the courts suggests that Congress should step in. To ensure accountability, and to encourage predictable outcomes, Congress should waive sovereign immunity for liability for all types of injuries and damages caused by negligently engineered, maintained, or operated structures or devices, whether they serve the purposes of "flood control," navigation, or something else entirely. At present, the Corps has little incentive to take even the most basic precautions required by the engineering profession, and injured parties are left on shaky ground.

(2) *Reform the decision-making metrics for construction of flood control projects.* The Flood Control Act of 1936 delegates broad discretion to the Corps to construct any flood control project it chooses. The only substantive constraint on the Corps' discretion is barely a constraint at all. The 1936 Act allows the Corps to proceed whenever "the benefits to whomsoever they may accrue are in excess of the estimated costs."

On its face, the requirement for a cost-benefit analysis (CBA) appears to provide a rational decision-making metric. By requiring that the costs and benefits of a proposed action be quantified and translated into dollars, CBA is said to be an unbiased method of evaluating the potential effects of a proposal and exposing bad plans that would impose ruinous costs. As applied, however, CBA requirements like this one are notoriously easy to manipulate, and excessive reliance on CBA has masked far too many bad policy choices. According to law professor Dan Tarlock, the Corps, in particular, "has a long history of inflated and methodologically unsound benefit-cost analysis techniques." Since 1936, the Corps has spent billions of dollars on dams, reservoirs, levees, and other structures, many of which have imposed social costs far in excess of the benefits produced. There are numerous examples of projects where the Corps overestimated project benefits and downplayed social and ecological costs in order to justify construction. According to some reports, the benefits of navigational manipulations on the Missouri to support virtually nonexistent commercial barge traffic were exaggerated nearly tenfold.[15] Likewise, in 2002, the National Research Council published a scathing critique of the Corps' use of inflated cost-benefit methodology to justify replacing old locks and dams on the upper Mississippi.

In the aftermath of the 2005 hurricanes, the Corps itself admitted that its approach to CBA is unlikely to justify risk reduction measures for storms like Katrina, in part because the analyses "do not consider such non-economic assets as human life."[16]

Cost-benefit analyses don't consider the value of human life?

How can this be good law? Congress must adopt more meaningful decision-making metrics. It should compel careful consideration of the value of human life and the potential for lost lives. Congress must also require the Corps to weigh the value of ecosystem services—buffering storm surges, capturing floodwaters, providing habitat for species, and furnishing recreational opportunities for humans—that may be lost as a result of flood control and related structures.

(3) *Reform the Fifth Amendment takings doctrine and empower governments to tame floods naturally by giving nature some space.* In its current form, the regulatory takings doctrine is unbalanced. Although it scrutinizes allegations of harm that property owners claim to suffer under "heavy-handed" laws, it does not call for a systematic inspection of the damage that these same landowners might otherwise inflict on their neighbors. Economists would call this a study of *externalities*, the spill-over effects that some landowners foist onto others. In lay terms, it comes down to being a good citizen and taking responsibility for the consequences of one's actions.

An article published in the *Yale Law Journal* developed the concept of *givings*—benefits conferred by the government that increase property values, such as the provision of flood control, subsidized federal insurance, sewers, roads, and the like. Before a court awards damages for a regulatory taking, the article suggests, judges should discount the claimed taking by the value added to the property by government action. This rough accounting of both give and take, the authors conclude, leads to a fairer, more balanced result.[17]

Whether we look at it as "takings" or "givings," if we expand our thinking about the relationship between landowners, their neighbors, governments, and the land itself, all sorts of possibilities may arise. Judges could incorporate several approaches to improve the doctrine, all of which recognize that landowners who misuse the land can "take" from their neighbors just as easily as government regulators can "take" from landowners.

First, giving greater weight to "ecosystem services" would quantify the benefits that landowners take from their neighbors when they use their property destructively (for example, by building on a wetland that could otherwise capture and tame floodwaters for the community).

In addition, a doctrine known as the *public trust* should be given greater force as a defense to alleged takings by the government. In *Just v. Marinette County*, the Wisconsin Supreme Court rejected a landowner's challenge to a county shoreline zoning ordinance that protected water quality by preventing landowners from changing the natural character of land adjacent to navigable rivers and lakes. The court held that the ordinance was not unreasonable, and that it "took" nothing from the landowner. The court observed that the state (and counties within the state) has an "active public trust duty . . . not only to promote navigation but also to protect and preserve [navigable] waters for fishing, recreation, and scenic beauty." The California Supreme Court followed suit in *National Audubon Society v. Superior Court,* where it concluded that the state Water Resources Control Board had a continuing duty to scrutinize Los Angeles' permit to use water from tributaries of Mono Lake. The court found that the Water Board was obligated to protect the public's interest in the beneficial use of its water, and to ensure against violations of the public trust by Los Angeles or other water users. As in *Just,* Los Angeles' assertion of a takings claim was rejected: The city had no vested property right to prevent the state from carrying out its public trust responsibilities.

Although the parameters of the public trust doctrine have not been clearly defined by the courts, the doctrine could be an especially powerful tool within the Mississippi River basin in light of the Northwest Ordinance of 1787, which provides that "The navigable waters leading into the Mississippi and St. Lawrence, and the carrying places between the same, shall be common highways and forever free . . . without any tax, impost, or duty thereof."[18] Instead of treating the ordinance as a relic of an earlier time, courts within the covered states—Illinois, Indiana, Michigan, Minnesota, Ohio, and Wisconsin—continue to cite it as proof of a pre-Constitutional, pre-statehood guarantee of public navigation, fishing, and recreation. Moreover, courts and scholars have recognized that the trust purposes are not set in stone, but can be expanded as the public's expectations expand. As the New Jersey Supreme Court stated, "The public trust doctrine . . . should not be considered fixed or static but should be molded and extended to meet changing conditions and needs of the public it was created to benefit."[19]

Surely, protecting people and ecosystems from improvident activities that destroy wetlands and shorelines, on the one hand, and preventing people and their structures from inflicting costs on their communities while placing themselves in harm's way in flood-prone areas, on the other, both come within the scope of the public trust doctrine. State and local governments should take action to prevent landowners from degrading public trust resources, and courts should protect those governments from "takings" claims when the landowners turn around and sue. As a result, it will be possible and perhaps even legally necessary to give nature some space to capture the natural bounty of its overflowing rivers.

(4) *Reform the National Flood Insurance Program to move people out of harm's way and to force those who engage in risky behavior to assume the risk.* This recommendation involves a number of related reform proposals.

(a) *Eliminate the levee loophole*: Before purchasing property, buyers (or their lenders) can measure the property's location against flood hazard maps prepared by the Federal Emergency Management Agency. The maps give would-be purchasers information about the potential flood risks they will face, and about whether or not they will be required to purchase federal flood insurance. Currently, if an otherwise flood-prone area lies behind a levee certified to protect from the 1 percent chance of an annual flood (the so-called *100-year flood*), the area can be removed from FEMA flood maps. As a consequence, landowners may have a false sense of security and not realize the risks they face if/when the levee fails. Congress should eliminate this levee loophole, which excludes certain flood-prone properties from the requirement to purchase insurance if the area lies behind a certified 100-year levee. At a minimum, Congress should limit the exclusion to properties behind 500-year levees. Even better, levees should not be an excuse for the failure to purchase flood insurance, if the area would otherwise be considered a floodplain. After all, we've learned there are only two kinds of levees: "Those that have failed and those that will fail!"

(*b*) *Require realistic federal maps and state disclosure forms*: Even if Congress does not repeal the levee loophole, states can take measures to improve landowners' understanding of risk. For example, state law could require sellers to disclose to potential purchasers the flood status of the property and the relevance of levees (if any) to that status. FEMA could require states that participate in the NFIP to adopt disclosure requirements. To assist sellers in satisfying their disclosure obligation, FEMA maps should be readily available and easy to read, and should make clear whether an area is outside the 100-year floodplain because of natural topography or, conversely, because of the presence of constructed levees.

(*c*) *Increase participation in the NFIP*: To make the program financially sustainable, more floodplain occupants must purchase and maintain insurance. First, officials must enforce existing purchase requirements. A 2006 study estimated that fewer than 50 percent of those living in hazardous areas had purchased mandatory flood insurance, and many who purchased insurance when they first got their mortgages subsequently dropped it. Second, Congress should expand the insurance requirement from 100-year floodplain occupants to 500-year floodplain occupants.

(*d*) *Adjust insurance premiums and eliminate subsidies*: Insurance premiums should not be subsidized at artificially low rates. This can be accomplished through four approaches:

1. Those who live behind levees should pay premiums that recognize the "residual risk" of levee failure.
2. Congress should adopt a "one strike and you're out" policy for repetitive risk properties. That is, floodplain residents would be compensated one time only. The payment should be sufficient to assist the resident in relocating to higher ground or in elevating the structure. If a second flood occurs, the resident will not be entitled to NFIP benefits.
3. Where older properties (generally less sturdy and less able to withstand flooding) have been "grandfathered" into the NFIP program at below-

market rates, Congress should phase out these subsidies over time to ensure that the premiums reflect the risks.

4. Congress should exclude voluntary risk-takers by making ineligible for NFIP coverage vacation homes and non-water dependent businesses, and by strengthening the exclusion of barrier island properties from NFIP.

(*e*) *End federal insurance altogether?* Bold reformers suggest eliminating the federal insurance program and allowing private insurers to step into the void (if they dare). This would likely either result in higher insurance premiums that more closely approximate the true cost of risky behavior, or eradicate coverage for flooding altogether. This proposal should be studied carefully to insure that it would discourage floodplain settlement without leaving stranded those with little choice. In addition, reformers must take care to ensure that one federal program (after-the-fact disaster relief) does not substitute for another (the NFIP).

* * *

OUR ATTITUDE TOWARD the Mississippi River is complicated, reflecting our feelings toward nature, in general. In *Last Child in the Wilderness: Saving Our Children from Nature-Deficit Disorder,* Richard Louv chronicles a shift in attitude over time. Following the so-called "frontier era," he says, we began unconsciously "[to] associate nature with doom —while disassociating the outdoors from joy and solitude." At the same time, he continues, "Americans romanticized, exploited, protected, and destroyed nature."[20] The stories in this book illustrate that difficult mix of impulses.

River romanticists cling to memories of an earlier time. Annie Gunn's Restaurant in the Missouri River floodplain idealizes the "richness of country life," even though the largest strip mall in the country, Chesterfield Commons, is right next door. The First English Evangelical Lutheran Church fought numerous court battles to retain its

Lutherglen campground in the Mill Creek floodplain above Los Angeles, California, even though floodwaters ravaged the site. Many states,
like Arkansas and Tennessee, have set the seemingly steadfast Mississippi as their boundary line. When the river shifts, as it inevitably does,
the states and the property owners who relied on clear title from the
states are left adrift.

We also exploit rivers, often with the best of intentions. Civic-minded
engineers like James Buchanan Eads squeezed the lower Mississippi
River between what he hoped would be an impervious line of levees to
protect the delta from floods. The Army Corps of Engineers created
the Mississippi River-Gulf Outlet (MRGO) as a short-cut to the Gulf
of Mexico in order to enhance the port of New Orleans, and the Corps
transformed the upper Mississippi into a navigable staircase of water
with twenty-nine pairs of locks and dams. The walled city of Cairo,
Illinois, took advantage of its enviable location near the confluence of
the Mississippi and Ohio rivers, but relied on an engineered floodway
to replace the natural floodplain. But sometimes we exploit nature on
senseless impulse, as suggested by the allegations against James Scott,
who was convicted of sabotaging the West Quincy levee during the
flood of 1993 in an ill-conceived plan to isolate his wife on the Missouri
side of the river and prevent her discovery of his mistress in Illinois.

We also try to protect nature through a series of laws and regulations.
The Clean Water Act establishes permit programs to protect wetlands
and to limit water pollution, and the Coastal Zone Management Act
offers protection to shorelines, dunes, and other sensitive areas.

In the end, though, many of these efforts have destroyed nature.
Levees prevent the lower Mississippi from nourishing its floodplain
soils with nutrients, and they restrict its delivery of land-building
sediments to the delta to help offset natural subsidence and erosion.
In the lower forty-eight states, over half of the original wetlands have
been filled, dredged, drained, or impaired. Iowa, Illinois, and Missouri have destroyed up to 85 percent of their pre-settlement wetlands.
In Louisiana, tens of thousands of miles of oil and gas pipelines and

canals transect the coastal marsh, allowing the intrusion of vegetation-killing saltwater.

*   *   *

BUT PAST NEED not be future. We have wasted more than a century pursuing a foolish ideal: *the floodless floodplain.* We struggled to hold back the water from the people. That effort failed time and again, as recounted by the flood stories in this book. We gave insufficient attention to the opposite notion—holding back the people from the waters. Despite recent experiments with taxpayer buy-outs of flood-prone properties, the old reliance on levees, insurance, and disaster relief continues to create irresistible incentives to settle in harm's way.

It's time to try something different: *giving rivers room to flood.* At the very least, we should think of sharing floodplains with their rivers. Like the Mississippi River itself, human attitudes shift over time as we build on experiences and learn from both success and failure. Attitudes take time to evolve. But as this book suggests, law can lead the way.

NOTES

NOTES TO THE INTRODUCTION

1. U.S. Army Corps of Engineers, New Orleans District, "Mississippi Flood Control: The Mississippi River."
2. U.S. Army Corps of Engineers, St. Paul District, "Regulating Mississippi River Navigation Pools."
3. U.S. Army Corps of Engineers, New Orleans District, "The Mississippi River & Tributaries (MR&T) Project."
4. U.S. Army Corps of Engineers, Headquarters, "Mission and Vision."
5. U.S. Army Corps of Engineers, Mississippi Valley Division, "Welcome."
6. Daniel A. Farber and Jim Chen, *Disasters and the Law: Katrina and Beyond*, xix–xx.
7. Ted Steinberg, *Acts of God: The Unnatural History of Natural Disaster in America*, xxi–xxii.
8. *Roath v. Driscoll*, 20 Conn. 533, 52 Am. Dec. 352 (1850).
9. *Tucker v. Badoian*, 376 Mass. 907, 384 N.E.2d 1195 (1978) (prospectively discarding common enemy rule in favor of reasonable use nuisance standard).
10. *Carland v. Aurin*, 103 Tenn. 555, 53 S.W. 940 (1899) (describing various legal theories).
11. *Armstrong v. Francis Corp.*, 20 N.J. 320, 120 A.2d 4 (1956).
12. *North v. Johnson*, 58 Minn. 242, 59 N.W. 1012 (1894).
13. Stephen Power, John Kell, and Siobhan Hughes, "BP, Oil Industry Take Fire at Hearing," *Wall Street Journal*, June 16, 2010.

NOTES TO CHAPTER 1

1. National Park Service, "Ice Age, National Scenic Trail."
2. National Park Service, "Explore Geology."
3. Bill Moyers Journal, "Losing Ground: The Disappearing Delta," September 6, 2002, quoting Tulane Law School Professor Oliver Houck's assertion that the Mississippi River delivers about a half million ton of silt to Louisiana each day, a load that would require 200,000 dump trucks daily. Estimates vary widely, however. See Louisiana Coastal Wetlands Planning, Protection & Restoration Act Program, "Mississippi River Basin Dynamics," estimating sediment load at 436,000 tons per day, or 160 million tons per year.

4. The Nature Conservancy, "The Upper Mississippi River Program."

5. John M. Barry, *Rising Tide: The Great Mississippi Flood of 1927 and How It Changed America*, 39.

6. NASA, Soil Science Education, "In the Time It Took to Form One Inch of Soil."

7. Langston Hughes, *The Negro Speaks of Rivers*.

8. Mark Twain, *Life on the Mississippi*, 25–26.

9. Nicola Cenacchi, *Assessing the Environmental Impact of Development Policy Lending on Coastal Areas: A World Bank Toolkit*, vii.

10. NOAA, "National Overview in Population Trends," 2005; NOAA, "Spatial Trends in Coastal Socioeconomics Demographic Trends Database," 2012.

11. Don Hinrichsen, "The Coastal Population Explosion," 29.

12. James Salzman, "Thirst: A Short History of Drinking Water," 99–100.

13. Barbara Kingsolver, "Fresh Water," 46.

14. Mark Twain, *Life on the Mississippi*, 2.

15. *Tennessee Statehood Act,* 1 Stat. 491 (1796); *Arkansas Statehood Act*, 5 Stat. 50 (1836).

16. University of Wisconsin-Stevens Point, "Oxbow Lake Formation," 2011.

17. *Cissna v. Tennessee*, 242 U.S. 195 (1916); *Cissna v. Tennessee*, 246 U.S. 289 (1918); *Arkansas v. Tennessee*, 246 U.S. 158, 173 (1918); *Arkansas v. Tennessee*, 247 U.S. 461 (1918); *Arkansas v. Tennessee*, 269 U.S. 152 (1925); *Arkansas v. Tennessee*, 397 U.S. 88 (1970); Twain, 242.

18. *Arkansas v. Tennessee*, 246 U.S. 158, 173 (1918).

19. "Island 37," *The Encyclopedia of Arkansas History and Culture*, 5081.

20. Ari Kelman, *A River and Its City: The Nature of Landscape in New Orleans*, 56–60.

21. William J. Petersen, *Steamboating on the Upper Mississippi*, 71.

22. Florence L. Dorsey, *Master of the Mississippi: Henry Shreve and the Conquest of the Mississippi* 101, 108, 118, 123–26.

23. Maurice G. Baxter, *The Steamboat Monopoly: Gibbons v. Ogden, 1824*, 31–36.

24. *Gibbons v. Ogden*, 22 U.S. 1 (1824).

25. *Louisiana Public Service Comm'n v. F.C.C.*, 476 U.S. 355 (1986).

26. A. A. Humphreys and Henry Abbot, *Report Upon the Physics and Hydraulics of the Mississippi River; Upon the Protection of the Alluvial Region Against Overflow; and Upon the Deepening of the Mouths*.

27. U.S. Army Corps of Engineers, "Federal Participation in Waterways Development"; John Barry, "After the Deluge."

28. Barry, *Rising Tide: The Great Mississippi Flood of 1927 and How It Changed America*, 47.

29. Ibid., 61.

30. Ibid., 75–76.

31. "Completion of the Jetties," *New York Times Tribune*, July 15, 1897.

32. Barry, *Rising Tide*, 75–76.
33. Beatrice Jones Hunter and Louis C. Hunter, *Steamboats on the Western Rivers: An Economic and Technological History*, 34.
34. Calvin R. Fremling, *Immortal River: The Upper Mississippi in Ancient and Modern Times*, 187–89.
35. Frederick Anderson et al., eds., *Mark Twain's Notebooks and Journals,* Volume III: 1883–1891, 529–30.
36. Mark Twain, *Life on the Mississippi*, 21.
37. U.S. Army Corps of Engineers, "Mississippi River Navigation."
38. Ibid.
39. "Mill City Museum," Minnesota Historical Society.
40. *Minnesota Chippewa Tribe v. U.S.*, 230 Ct.Cl. 776 (1982).
41. "Historic American Engineering Record, Lake Winnibigoshish Reservoir Dam, Deer River Vicinity, Itasca County, MN, Survey No. HAER MN-65," Library of Congress.
42. John O. Anfinson, *The River We Have Wrought: A History of the Upper Mississippi*, 90.
43. Fremling, *Immortal River*, 206.
44. Ibid.
45. *Rivers and Harbors Act of 1907*, ch. 2509, 34 Stat. 1110 (March 2, 1907).
46. Clayton D. Brown, "Western Tributaries of the Mississippi," 7.
47. *Rivers and Harbors Act of 1930*, ch. 847, 46 Stat. 918, 927, 928 (July 3, 1930).
48. U.S. Geological Survey, "About the Upper Mississippi River System."
49. U.S. Army Corps of Engineers, New Orleans District, "The Mississippi River and Tributaries Project."
50. The Nature Conservancy, "The Upper Mississippi River Program."
51. Twain, 301–302.
52. U.S. Army Corps of Engineers, New Orleans District, "The Mississippi Drainage Basin."

NOTES TO CHAPTER 2

1. Garcilaso de la Vega, *The Florida of the Inca: A History of Adelantado, Hernando de Soto, Governor and Captain General of the Kingdom of Florida, and of Other Heroic Spanish and Indian Cavaliers*, 552–54.
2. Ibid., 535.
3. Tristram R. Kidder, "Making the City Inevitable: Native Americans and the Geography of New Orleans," 13.
4. Fremling, 116.
5. Barry, *Rising Tide*, 34.
6. D. Gorton, "Methods and Materials [of J.C. Coovert]."
7. W. J. McGee, "The Flood Plains of Rivers," 221–22.
8. Ibid., 225.

9. National Research Council, *Mississippi River Water Quality and the Clean Water Act: Progress, Challenges, and Opportunities*, 139.

10. Lanna Combs & Charles Perry, "The 1903 and 1993 Floods in Kansas—The Effects of Changing Times and Technology," 1.

11. Harry Crawford Frankenfield, *The Floods of the Spring of 1903, in the Mississippi Watershed*, 29.

12. "New Orleans Levee Systems: Hurricane Katrina," Independent Levee Investigation Team.

13. Frankenfield, 29.

14. "Harlem, MO Flood, May 1903."

15. Nikaela J. Zimmerman, "The Flood-1903: Cafe Clock."

16. Robert Kelley Schneiders, *Unruly River: Two Centuries of Change along the Missouri*, 85.

17. *Moffatt Commission Co. v. Union Pac. R. Co.*, 113 Mo.App. 544, 88 S.W. 117 (1905).

18. Ibid., 118.

19. *Empire State Cattle Co. v. Atchison, T. & S. F. Ry. Co.*, 135 F.135 (D. Kan. 1905), aff'd 210 U.S. 1 (1908).

20. R. G. Dun & Company, *Dun's Review*, 40.

21. *Empire State Cattle Co. v. Atchison, T. & S. F. Ry. Co.*, 143–44.

22. James Willard Hurst, *Law and Economic Growth: The Legal History of the Lumber Industry in Wisconsin, 1836–1915*; Richard A. Posner, "A Theory of Negligence," 1.

23. *United States v. Carroll Towing Co.*, 159 F.2d 169 (2d Cir. 1947); *Palsgraf v. Long Island R.R. Co.*, 162 N.E. 99 (N.Y. 1928).

24. Iain White, "The Absorbent City: Urban Form and Flood Risk Management," 152; Gilbert F. White, "Human Adjustment to Floods," 29.

25. Anfinson, 98.

26. *Gibbons v. Ogden*.

27. Donald Pisani, *Water and American Government: The Reclamation Bureau, National Water Policy, and the West*, 253.

28. George Fitch, "The Missouri River: Its Habits and Eccentricities Described by a Personal Friend," 637–40.

29. Schneiders, 98.

30. Brown, "The Mississippi River Flood of 1912," 645–57.

31. J. W. Grubaugh and R. V. Anderson, "Long Term Effects of Navigation Dams on a Segment of the Upper Mississippi River," 102.

32. John Solomon Otto, *The Final Frontiers, 1880–1930: Settling the Southern Bottomlands*, 49.

33. Ibid.

34. "Human Dike Used to Hold Back Flood," *New York Times*, April 11, 1912.

35. Terry L. Jones, "Looming Menace?"

36. "Citizens Fight Flood Along with Convicts; Rich and Poor at Baton Rouge Unite in Desperate Struggle," *New York Times*, May 4, 1912.

37. "Fresh Perils Added to Flood by Storm; Severest Rain in History of New Orleans Renews Danger in Inundated Southern Territory," *New York Times*, May 11, 1912.

38. J. Y. Sanders, "Governor Sanders Tells of Big Flood Losses; 375,000 Persons Homeless and Property Damage $6,000,000," *New York Times,* May 8,1912.

39. Brown, "The Mississippi River Flood of 1912," 646, 651, 653.

40. "Denies 300 Drowned in Southern Flood; Gov. Brewer of Mississippi Telegraphs The Times That Report Is Untrue," *New York Times*, April 21, 1912.

41. "The Virginia Newspaper Project Examines the News Covering the Sinking of R.M.S. *Titanic*, April 14, 1912," Library of Virginia.

42. Ibid.

43. Grubaugh and Anderson, 102.

44. Brown, "The Mississippi River Flood of 1912," 655.

45. Ohio Historical Society, "Severe Weather in Ohio."

46. U.S. Department of Agriculture, Weather Bureau, *The Floods of 1913 in the Rivers of the Ohio and Lower Mississippi Valleys*, 30.

47. "Memphis in Grave Peril; Levee Near Wilson, Ark., Goes Out—Other Levees Menaced," *New York Times*, April 9, 1913.

48. "New Orleans Raises Dikes; Expects to Have Them Well Above Mississippi's Flood Crest," *New York Times*, April 12, 1913.

49. "Memphis in Grave Peril; Levee Near Wilson, Ark., Goes Out—Other Levees Menaced," *New York Times,* April 9, 1913.

50. 1913 Flood Postcard Collection, Dayton, Ohio, Metro Library.

51. *Watts v. Evansville, Mt. C. & N. Ry. Co.*, 120 N.E. 613 (Ind. App. 2 Div. 1918).

52. Ibid., 129 N.E. 322–24 (Ind. 1921).

53. Jeannie Whayne, "Robert Edward Lee Wilson (1865–1933)."

54. *Act of June 28, 1879*, 21 Stat. 37; *Jackson v. U.S.*, 230 U.S. 16 (1913).

55. *Jackson v. U.S.*, 230 U.S. 19 (1913).

56. W. M. Black, "The Problem of the Mississippi," 636–42; Martin Reuss, "Andrew A. Humphreys and the Development of Hydraulic Engineering: Politics and Technology in the Army Corps of Engineers, 1850–1950," 27, 32–33.

57. Samuel P. Hays, *Conservation and the Gospel of Efficiency: The Progressive Conservation Movement, 1890–1920*, 210–18.

58. Edward Marshall, "Willcocks Sees Simple Solution of Mississippi Flood Problem; Sir William Willcocks, Designer of the Assouan Dam, Says Our Problem Is Easier Than That of the Mesopotamian Rivers, Solved 5,000 Years Ago," *New York Times,* SM7.

59. Ibid.

60. Ibid.

61. *Act of Mar. 1, 1917*, 64th Cong., Pub. Law No. 367.

62. *Cubbins v. The Mississippi River Commission*, 241 U.S. 351 (1916).

63. *U.S. v. Archer*, 241 U.S. 129 (1916).

64. Matthew T. Pearcy, "A History of the Ransdell-Humphreys Flood Control Act of 1917," 133–59.

65. U.S. Fish & Wildlife Service, "Holt Collier National Wildlife Refuge: Refuge History."

NOTES TO CHAPTER 3

1. George Brown Tindall, *The Emergence of the New South, 1913–1945*, 111, 121.

2. Claude S. Fischer, "Technology's Retreat: The Decline of Rural Telephony in the United States, 1920–1940," 3.

3. Charles Angoff, *H. L. Mencken: A Portrait from Memory*, 126.

4. Jonathan Yardley, "The Sage of Baltimore," *Atlantic Monthly*, 129–40.

5. Tindall, 211.

6. *Mammoth Oil Company v. United States*, 275 U.S. 13 (1927).

7. Robert W. Cherny, "Graft and Oil: How Teapot Dome Became the Greatest Political Scandal of its Time."

8. Charles J. Shindo, *1927 and the Rise of Modern America*, 1–2, 7–8; Gerald Leinwand, *1927: High Tide of the Twenties.*

9. Tindall, 83.

10. Harry Taylor, "Problems of Flood Control," 345.

11. Donald Pisani, *Water and American Government: The Reclamation Bureau, National Water Policy, and the West*, 235.

12. Ibid., 253; U.S. Chamber of Commerce, 7 Cong. Digest 44 (1928).

13. *Act of Mar. 1, 1917*, 64th Cong., Pub. Law No. 367.

14. U.S. Chamber of Commerce, 7 Cong. Digest 45 (1928).

15. Pisani, 253.

16. Ibid., 249.

17. Pete Daniel, *Deep'n As It Come: The 1927 Mississippi River Flood*, 6; Annual Report of the Chief of Engineers, U.S. Army, *Mississippi River Commission* (1926), 1793.

18. Barry, "After the Deluge," 114.

19. U.S. Army Corps of Engineers, Rock Island District, "Levee Safety Program: Levee Terms & Definitions."

20. Barry, *Rising Tide*, 196.

21. Pisani, 250.

22. Barry, *Rising Tide*, 312–13.

23. Ibid., 200–202.

24. "Peonage Bill Hits Workers in Louisiana: Plantation Laborers Are Made Serfs," *Chicago Defender*, 1; Gabriel J. Chin, "The Jena Six and the History of Racially Compromised Justice in Louisiana," 376; Jennifer Roback, "Southern Labor Law in the Jim Crow Era: Exploitative or Competitive?" 1169.

25. Barry, *Rising Tide*, 200–202, quoting *Memphis Commercial-Appeal* (1927).

26. Barry, "After the Deluge," 116.

27. International Communist League, "Black Oppression and the Great Mississippi Flood of 1927."

28. Vincent Fitzpatrick, *H. L. Mencken*, 66.

29. Bearden, 7–8.

30. *United States v. James*, 478 U.S. 597 (1986); Federal Reserve Bank of Minneapolis, "What Is A Dollar Worth?"

31. Russell E. Bearden, "Arkansas' Worst Disaster: The Great Mississippi River Flood of 1927," 7–8.

32. Karen M. O'Neill, *Rivers by Design: State Power and the Origins of U.S. Flood Control*, 140.

33. Bearden, 6.

34. Barry, "After the Deluge," 115.

35. Pisani, 250.

36. Barry, *Rising Tide*, 253–58.

37. *Foret v. Board of Levee Com'rs of Orleans Levee Dist.*, 169 La. 427, 125 So. 437 (1929).

38. *Alfred Oliver & Co. v. Board of Com'rs of Orleans Levee Dist.*, 169 La. 438, 125 So. 441 (1929).

39. Glen Jeansonne, *Leander Perez: Boss of the Delta*; William Ivy Hair, *The King-fish and His Realm: The Life and Times of Huey P. Long*.

40. Barry, *Rising Tide*, 407.

41. U.S. Chamber of Commerce, *Congressional Digest* 7 (1928).

42. *United States v. James* (1985); 69 Cong. Rec. 5294 (1928).

43. Pisani, 251–52; "Control of Floods on Mississippi River," S. Rep. No. 70-448 (1928).

44. *United States v. James*, 760 F.2d 590 (5th Cir. 1985); 69 Cong.Rec. 5294 (1928).

45. 69 Cong. Rec. 7011 (1928).

46. O'Neill, 144.

47. Barry, "After the Deluge," 115.

48. U.S. Army Corps of Engineers, Vicksburg District, "A Century of Service."

49. U. S. Chamber of Commerce, *Congressional Digest*, 46.

50. Ibid.

51. Barry, *Rising Tide*, 406.

52. *Flood Control Act of 1928*, 45 Stat. 534, 33 U.S.C. §§ 702a-m.

53. Barry, "After the Deluge," 116.

54. O'Neill, 146.

55. Ibid.

56. *Flood Control Act of 1928*, 33 U.S.C. § 701.

57. Ibid., § 702.

58. *Marsh v. Oregon Natural Resources Council*, 490 U.S. 360 (1989); *In re Operation of Missouri River System Litigation*, 421 F.3d 618 (8th Cir. 2005).

59. H.R. Rep. No. 1101, 70th Cong., 1st Sess. 13 (1928).

NOTES TO CHAPTER 4

1. Alice Talmadge, "Dust, Drought, and Despair."

2. Andrew B. Abel, Ben S. Bernanke, and Dean Croushore, *Macroeconomics*, 285–86.

3. Timothy Egan, *The Worst Hard Time: The Untold Story of Those Who Survived the Great American Dust Bowl*, 10, 122.

4. *Soil Conservation Act of 1935*, Pub. L. No. 74-46, 49 Stat. 163 (codified as amended at 16 U.S.C. §§ 590a-590-3).

5. William J. Futrell, "The IUCN Sustainable Soil Project and Enforcement Failures," 103–104.

6. Sandra Zellmer, "The Devil, the Details, and the Dawn of the 21st Century Administrative State: Beyond the New Deal," 941.

7. Pisani, 271.

8. *Flood Control Act of 1936*, 33 U.S.C. § 701a.

9. Ibid.

10. Christine A. Klein, "On Dams and Democracy," 641; Marc Reisner, *Cadillac Desert: The American West and its Disappearing Water*, 178.

11. John M. Lansden, *A History of the City of Cairo, Illinois*, 252–54 (1st ed. 2009).

12. Annual Report of the Chief of Engineers, U.S. Army, *Mississippi River Commission*, 4.

13. Ibid., 5.

14. Ibid., 7.

15. NOAA, National Weather Service Weather Forecast Service, "The Great Flood of 1937."

16. Jarod Roll, "Out Yonder on the Road: Working Class Self-Representation and the 1939 Roadside Demonstration in Southeast Missouri."

17. U.S. Army Corps of Engineers, New Orleans District, "Mississippi Flood Control: The Mississippi River."

18. Annual Report of the Chief of Engineers, *Mississippi River Commission*, 8.

19. NOAA, National Weather Service Weather Forecast Office, "The Great Flood of 1937."

20. Aaron Lake Smith, "Trying to Revitalize a Dying Small Town."

21. Mark Guarino, "Mississippi River Flooding: After Levee Blast, Threat Shifts

to Memphis"; David Mercer, "Cairo Evacuated: Mayor Tells Residents to Leave Flood-Threatened Illinois Town"; A.G. Sulzberger, "Army Corps Blows Up Missouri Levee."

22. *Big Oak Farms, Inc. v. United States*, 105 Fed. Cl. 48 (2012).

NOTES TO CHAPTER 5

1. Rena I. Steinzor, "Unfunded Environmental Mandates and the New (New) Federalism: Devolution, Revolution, or Reform?" 116.
2. *Flood Control Act of 1944*, 33 U.S.C. 701-1.
3. Ibid.
4. Barbara Burgess, "Topeka's Roots: The Prairie Potato."
5. National Museum of American History, "Segregation in the Heartland."
6. U.S. Geological Survey, "Kansas Floods."
7. NET Nebraska, "1952 Missouri River Flood: A Look Back."
8. NOAA, "Historic Flood Events in the Missouri River Basin."
9. National Research Council, *Missouri River Planning: Recognizing and Incorporating Sediment Management,* 78–83.
10. Michael L. Lawson, *Dammed Indians: The Pick-Sloan Plan and the Missouri River Sioux*, 23.
11. National Research Council, *Missouri River Planning*, 31.
12. Lawson, *Dammed Indians*, 25.
13. National Research Council, *Missouri River Planning*, 31.
14. Lawson, *Dammed Indians*, 30.
15. Ibid., 33.
16. Ibid., 26.
17. Mark Michel, "The Archaeological Conservancy and Site Protection."
18. Gabrielle Paschall, "Protecting Our Past: The Need for Uniform Regulation to Protect Archeological Resources," 372.
19. Derek V. Goodwin, "Raiders of the Sacred Sites."
20. Jori Finkel, "Is Everything Sacred? A Respected Art Dealer is Busted for Selling a Cheyenne War Bonnet," 65–66.
21. *Yankton Sioux Tribe v. U.S. Army Corps of Eng'rs*, 83 F. Supp. 2d 1047 (D.S.D.2000).
22. Ibid.

NOTES TO CHAPTER 6

1. John McPhee, "Atchafalaya," *New Yorker*, February 23, 1987, 5.
2. Thomas A. Lewis, *Brace for Impact: Surviving the Crash of the Industrial Age by Sustainable Living*, 188.
3. Ibid. (transcription of Army Corps of Engineers promotional film).
4. U.S. Army Corps of Engineers, "Old River Control." The Center for Land Use Interpretation, "Old River Control Structure."

5. McPhee, "Atchafalaya."
6. Lewis, *Brace for Impact*.
7. Peirce F. Lewis, *New Orleans: The Making of an Urban Landscape*, 17.
8. Christopher Morris, "Impenetrable but Easy," 28, 31.
9. Ivor Van Heerden and Mike Bryan, *The Storm: What Went Wrong and Why During Hurricane Katrina—The Inside Story from One Louisiana Scientist*, 158–70.
10. Craig E. Colten, *Perilous Place, Powerful Storms: Hurricane Protection in Coastal Louisiana*, 89.
11. Kelman, xiv–xv.
12. U.S. Army Corps of Engineers, Team New Orleans, "Spillway Operation Information."
13. *Flood Control Act of 1928, 33* U.S.C. § 702.
14. *U.S. v. James*, 478 U.S. 597 (1986).
15. *Fryman v. United States*, 901 F.2d 79 (7th Cir. 1990).
16. *Central Green Co. v. United States*, 531 U.S. 425 (2001) (abrogating *James*, 1986).
17. U.S. Army Corps of Engineers, "Information About MRGO."
18. Todd Shallat, "In the Wake of Hurricane Betsy," 128–30.
19. Ibid., 122, quoting from "Hearings before the Special Subcommittee to Investigate Areas of Destruction of Hurricane Betsy."
20. Jerry D. Jarrell, Max Mayfield, and Edward N. Rappaport, *The Deadliest, Costliest, and Most Intense United States Hurricanes From 1900 to 2000*.
21. *Graci v. United States*, 301 F. Supp. 947 (E.D. La. 1969).
22. Ibid., 456 F.2d 25 (5th Cir. 1971).
23. Ibid., 456 F.2d 27 (5th Cir. 1971).
24. Ibid., 456 F.2d 25 (5th Cir. 1971).
25. Ibid., 435 F. Supp. 189 (E.D. La. 1977).
26. Oliver A. Houck, "Rising Water: The National Flood Insurance Program and Louisiana," 67; Saul Jay Singer, "Flooding the Fifth Amendment: The National Flood Insurance Program and the Takings Clause," 334.
27. Pub. L. No. 84-1016, 70 Stat. 1078 (1956) (repealed 1957).
28. *Senate Report No. 93-583* (1973).
29. *Stafford Disaster Relief and Emergency Assistance Act*, Pub.L. 100-707, Title I, § 105(c), Nov. 23, 1988, 102 Stat. 4691.
30. *First English Evangelical Lutheran Church of Glendale v. County of Los Angeles* (Brief for Appellant), 2–4.
31. *First English Evangelical Lutheran Church of Glendale v. County of Los Angeles*, 482 U.S. 318-19 (1987).
32. Ibid., 482 U.S. 340-44 (1987).
33. *Adolph v. Federal Emergency Management Agency*, 854 F.2d 732 (5th Cir. 1988).

34. Vicki Been, "*Lucas v. The Green Machine*: Using the Takings Clause to Promote More Efficient Regulation?" 225.

35. Dukeminier, Jesse, James E. Krier, Gregory S. Alexander, and Michael H. Schill, *Property*, 1150.

36. South Carolina State Climatology Office, "South Carolina Hurricane Climatology—Notable South Carolina Hurricanes."

37. *Lucas v. S. Carolina Coastal Council*, 505 U.S. 1003, 1070 (1992).

38. General Accounting Office, "Army Corps of Engineers: Lake Pontchartrain and Vicinity Hurricane Protection Project," Testimony of Anu Mittal Before the Subcommittee on Energy and Water Development, Committee on Appropriations, House of Representatives.

NOTES TO CHAPTER 7

1. Water Science & Technology Board, Commission on Flood Control Alternatives in the American River Basin, "Flood Risk Management and the American River Basin: An Evaluation," 164.

2. Steinberg, 98.

3. Missouri State Emergency Management Agency, "10-Year Anniversary."

4. Isabel Wilkerson, "Cruel Flood: It Tore at Graves, and at Hearts," 41; Rutherford H. Platt, "Review of Sharing the Challenge: Floodplain Management into the 21st Century," 26.

5. Wilkerson, 41; Platt, 26.

6. Wilkerson, 41.

7. *Missouri v. Scott*, 943 S.W.2d 730, 733 (Mo. App. 1997) (citing Missouri Revised Statutes § 569.070).

8. Ibid.

9. Adam Pitluk, "Dammed to Eternity."

10. 44 Code of Federal Regulations 59.2 (2006).

11. *National Flood Insurance Reform Act of 1994*, 108 Stat. 2255 (codified as amended in scattered sections of 42 U.S. Code).

NOTES TO CHAPTER 8

1. U.S. Geological Survey, "About Wetlands."

2. *Leovy v. United States*, 177 U.S. 621 (1900).

3. *Sabine River Authority v. U.S. Department of the Interior*, 951 F.2d 669 (5th Cir. 1992).

4. *Allison v. Barberry Homes, Inc.*, No. 982935, WL 1473121 (2000).

5. *Hoffman Homes, Inc. v. EPA*, 999 F.2d 262 (1993).

6. U.S. Environmental Protection Agency, "Economic Benefits of Wetlands."

7. U.S. Environmental Protection Agency, "Watershed Academy: Wetland Functions and Values," 7–8.

8. Bob Sullivan, "Wetlands Erosion Raises Hurricane Risks."

9. U.S. Environmental Protection Agency, "Watershed Academy," 6.
10. Ibid.
11. Ibid., citing 1997 data from the Pacific Coast Federation of Fishermen's Associations.
12. Ibid., 10–11.
13. Louisiana Coastal Wetlands Planning Protection & Restoration Act Program, "Mississippi River Basin Dynamics."
14. America's Wetland Foundation, "About Us."
15. U.S. Geological Survey, "Louisiana Water."
16. John Barry, "Battling Nature on the River," C1.
17. John McQuaid and Mark Schleifstein, *In Harm's Way: Surging Water Is the Biggest Threat to New Orleans*, (New Orleans) *Times-Picayune*, J2.
18. McQuaid & Schleifstein, *In Harm's Way*.
19. Louisiana Coastal Wetlands Planning Protection & Restoration Act Program, "Mississippi River Basin Dynamics."
20. David M. Driesen et al., "An Unnatural Disaster: The Aftermath of Hurricane Katrina," 13–14.
21. General Accounting Office, "Army Corps of Engineers: Lake Pontchartrain and Vicinity Hurricane Protection Project," 7.
22. John McPhee, *The Control of Nature*, 59.
23. Leah Hodges, "New Orleans Evacuees and Activists Testify at Explosive House Hearing on the Role of Race and Class in Government's Response to Hurricane Katrina."
24. Michael Fletcher and Richard Morin, "Bush's Approval Rating Drops to New Low in Wake of Storm," A08.
25. Steve Inskeep and David Greene, "President Bush Returns to New Orleans."
26. Juliana Maantay & Andrew Maroko, "Mapping Urban Risk."
27. Michelle Krupa, "Ethel Williams Dies at 75: 9th Ward Resident Took Bush to Flooded Home."
28. Sharon Begley, "Man-Made Mistakes Increase Devastation of 'Natural' Disasters."
29. U.S. Army Corps of Engineers, "Performance Evaluation of the New Orleans and Southeast Louisiana Hurricane Protection System."
30. *In re Katrina Canal Breaches Consolidated Litigation*, 533 F.Supp.2d 615 (E.D. La. 2008).
31. Ibid.
32. *In re Katrina Canal Breaches Consolidated Litigation*, 647 F.Supp.2d 644 (E.D. La. 2009).
33. *In re Katrina Canal Breaches Consolidated Litigation*, 673 F.3d 381 (5th Cir. 2012), *withdrawn on rehearing*, 696 F.3d 436 (5th Cir. 2012), *cert. denied*, 133 S.Ct. 2855 (2013).
34. *Public Law No. 109-234.*

35. Lewis, *New Orleans: The Making of an Urban Landscape*, 17.
36. Kelman, xiii–xv.

NOTES TO CHAPTER 9

1. Richard Wright, *Uncle Tom's Children*, 97.
2. PETE: National Partnership for Environmental Technology Education, "NSF Tribal."
3. United Church of Christ, Commission for Racial Justice, *Toxic Waste and Race in the United States: A National Report on the Racial and Socio-Economic Characteristics of Communities with Hazardous Waste Sites.*
4. Eileen McGurty, *Transforming Environmentalism: Warren County, PCBs, and the Origins of Environmental Justice*, 167; Robert D. Bullard, *Dumping in Dixie: Race, Class and Environmental Quality.*
5. 15 U.S.C. § 2605(e); 40 C.F.R. Part 761.
6. *United States v. Ward*, 676 F.2d 94 (4th Cir. 1982).
7. Benjamin Chavis, "Concerning the Historical Evolution of the 'Environmental Justice Movement' and the Definition of the Term: 'Environmental Racism.'"
8. *Civil Rights Act of 1964*, 42 U.S.C. § 2000d.
9. *NAACP v. Gorsuch*, No. 82-768-CIV-5, slip op. at 10 (E.D.N.C. Aug. 1982).
10. Bradford Mank, "Title VI and the Warren County Protests," 74.
11. *Alexander v. Sandoval*, 532 U.S. 275 (2001).
12. McGurty, *Transforming Environmentalism*, 301–323.
13. Robert D. Bullard and Beverly Wright, "Disastrous Response to Natural and Man-Made Disasters: An Environmental Justice Analysis Twenty-Five Years After Warren County," 224.
14. UNC Exchange Project, "Real People-Real Stories: Seeking Environmental Justice."
15. Bullard and Wright, 222.
16. Warren County, "Warren County-2011, Comprehensive Development Plan," 77.
17. Luci Weldon, "State Deeds PCB Landfill to County."
18. Elana Schor, "Superfund: At Midlife, Unending Cleanups and Less 'Real Money'; Sandra George O'Neil, "Superfund: Evaluating the Impact of Executive Order 12898."
19. Susan L. Cutter, *Hazards, Vulnerability and Environmental Justice*, xxi–xxiii.
20. Craig E. Colten, "Environmental Justice on the American Bottom."
21. A. Fothergill, E. G. M. Maestas, and J. D. Darlington, "Race, Ethnicity, and Disasters in the United States: A Review of the Literature," 165.
22. Cutter, xxi–xxiii.
23. Mike Tidwell, *The Ravaging Tide: Strange Weather, Future Katrinas, and The Coming Death of America's Coastal Cities*, 27.
24. Fothergill, Maestas, and Darlington, 165.

25. Juliana Maantay and Andrew Maroko, "Mapping Urban Risk: Flood Hazards, Race, & Environmental Justice in New York," 111–12.
26. Jared Diamond, *Collapse: How Societies Choose to Fail or Succeed*, 160–77.
27. Jared Diamond, "The Last Americans, Environmental Collapse and the End of Civilization," 43–51.
28. "Why Societies Collapse: Jared Diamond at Princeton University," Australian Broadcasting Corporation transcript.
29. U.S. Environmental Protection Agency, *Environmental Equity: Reducing Risk for All Communities*, 3.
30. Robert D. Bullard, *The Quest for Environmental Justice*, 1–15.
31. Exec. Order No. 12,898 (1994).
32. Ibid.
33. Bullard and Wright, 225.

NOTES TO CHAPTER 10

1. Peter J. Byrne, "Ten Arguments for the Abolition of the Regulatory Takings Doctrine," 102; Carol M. Rose, "Mahon Reconstructed: Why the Takings Doctrine Is Still a Muddle," 561.
2. James E. Krier, "The Takings-Puzzle Puzzle."
3. Louise A. Halper, "Why the Nuisance Knot Can't Undo the Takings Muddle," 329.
4. *Lingle v. Chevron U.S.A., Inc.*, 544 U.S. 528, 540 (transcript of oral argument, 21).
5. *Pennsylvania Central Transportation Company v. City of New York*, 438 U.S. 104 (1978).
6. Hearings Before the Senate Committee on Environmental & Public Works, "Comprehensive and Integrated Approach to Meet the Water Resources Needs in the Wake of Hurricanes Katrina and Rita," statement of Scott Faber.
7. *Gove v. Zoning Board of Appeals of Chatham*, 831 N.E.2d 865 (Mass. 2005).
8. *Adolph v. Federal Emergency Management Agency*, 854 F.2d 732 (5th Cir. 1988).

NOTES TO THE CONCLUSION

1. Richard Louv, *Last Child in the Woods: Saving Our Children from Nature-Deficit Disorder*, 159.
2. Charles A. Perry, "Significant Floods in the United States during the 20th Century: USGS Measures a Century of Floods."
3. National Oceanic and Atmospheric Administration, "United States Flood Loss Report—Water Year 2011," 2–6.
4. Bryan Martinsdale and Paul Osman, Illinois Association for Floodplain and Stormwater, "Why the Concerns with Levees? They're Safe, Right?"

5. American Society of Civil Engineers, "So You Live Behind a Levee! What You Should Know to Protect Your Home and Loved Ones from Floods."

6. Saul Jay Singer, "Flooding the Fifth Amendment: The National Flood Insurance Program and the Takings Clause," 334.

7. *Flood Disaster Protection Act of 1973* (establishing mandatory flood insurance purchase requirement for structures in areas identified as SFHA, enforced through loans by federally-insured lenders).

8. *National Flood Insurance Reform Act of 1994* (strengthening lender compliance with mandatory purchase requirements).

9. *Flood Insurance Reform Act of 2004* (establishing pilot program for severe repetitive loss properties through mitigation).

10. Rawle O. King, "National Flood Insurance Program: Background, Challenges, and Financial Status," 1.

11. Ibid., 4, citing data from the U.S. Department of Homeland Security, Federal Emergency Management Agency.

12. Ibid., 1.

13. Rawle O. King, "Federal Flood Insurance: The Repetitive Loss Problem," 19–20, 27.

14. Nicole Norfleet, "Melting Snow Creates Anxiety at C&O Canal's Potomac Gate."

15. Bruce Upin, "A River of Subsidies," 86.

16. Mark Schleifstein, "Corps Report Ignores Call for Specifics; Details for Category 5 Protections Left Out."

17. Abraham Bell and Gideon Parchomovsky, "Givings," 547.

18. *Ordinance of the Northwest Territory*, section 14, article IV, July 13, 1787.

19. *Slocum v. Borough of Belmar*, 569 A.2d 312, 316 (N.J. Super. L. 1989).

20. Louv, *Last Child in the Woods*, 4.

BIBLIOGRAPHY

BOOKS

Abel, Andrew B., Ben S. Bernanke, and Dean Croushore. *Macroeconomics*, 6th ed. Boston: Pearson Education, Addison-Wesley, 2008.

Ambrose, Stephen E. *Undaunted Courage: Meriwether Lewis, Thomas Jefferson, and the Opening of the American West*. New York: Simon and Schuster, 2003.

Anderson, Frederick, Lin Salamo, Michael B. Frank, and Robert Pack Browning, eds. *Mark Twain's Notebooks and Journals*, vol. 3: 1883–1891. Berkeley: University of California Press, 1980.

Anfinson, John O. *The River We Have Wrought: A History of the Upper Mississippi*. Minneapolis: University of Minnesota Press, 2005.

Angoff, Charles. *H. L. Mencken: A Portrait from Memory*. New York: T. Yoseloff, 1956.

Barry, John M. *Rising Tide: The Great Mississippi Flood of 1927 and How it Changed America*, Touchstone ed. New York: Simon and Schuster, 1998.

Bates, Sarah F., et al. *Searching Out the Headwaters: Change and Rediscovery in Western Water Policy*. Washington DC: Island Press, 1993.

Baxter, Maurice G. *The Steamboat Monopoly: Gibbons v. Ogden, 1824*. New York: Knopf Publishing Group, 1972.

Bullard, Robert D. *Dumping in Dixie: Race, Class and Environmental Quality*. Boulder, CO: Westview Press, 2000.

———. *The Quest for Environmental Justice: Human Rights and the Politics of Pollution*. San Francisco: Sierra Club Books, 2005.

———, ed. *Unequal Protection: Environmental Justice and Communities of Color*. San Francisco: Sierra Club Books, 1997.

Carrels, Peter. *Uphill against Water: The Great Dakota Water War*. Lincoln: University of Nebraska Press, 1999.

Cenacchi, Nicola. *Assessing the Environmental Impact of Development Policy Lending on Coastal Areas: A World Bank Toolkit*. Washington, DC: World Bank, 2010.

Colten, Craig E. *Perilous Place, Powerful Storms: Hurricane Protection in Coastal Louisiana*. Jackson: University Press of Mississippi, 2009.

———. *Transforming New Orleans and Its Environs: Centuries of Change*. Pittsburgh: University of Pittsburgh Press, 2000.

Cutter, Susan L. *Hazards, Vulnerability and Environmental Justice.* Sterling: Routledge Earthscan, 2006.

Daniel, Pete. *Deep'n As It Come: The 1927 Mississippi River Flood.* New York: Oxford University Press, 1977.

De La Vega, Garcilaso. *The Florida of the Inca: A History of Adelantado, Hernando de Soto, Governor and Captain General of the Kingdom of Florida, and of Other Heroic Spanish and Indian Cavaliers.* Austin: University of Texas Press, 1951.

Diamond, Jared. *Collapse: How Societies Choose to Fail or Succeed.* New York: Penguin Books, 2004.

Dorsey, Florence L. *Master of the Mississippi: Henry Shreve and the Conquest of the Mississippi.* Gretna, LA: Pelican Publishing, 1998.

Dukeminier, Jesse, James E. Krier, Gregory S. Alexander, and Michael H. Schill. *Property*, 7th ed. New York: Aspen Publishers, 2010.

Dworzak, Thomas, Stanley Greene, Kadir van Lohuizen, and Paolo Pellegrin. *Katrina: An Unnatural Disaster.* Santa Monica, CA: Ram Distribution, 2006.

Egan, Timothy. *The Worst Hard Time: The Untold Story of Those Who Survived the Great American Dust Bowl.* New York: Houghton Mifflin, 2006.

Farber, Daniel A., and Jim Chen. *Disasters and the Law: Katrina and Beyond.* New York: Aspen Publishers, 2006.

Fitzpatrick, Vincent. *H. L. Mencken.* Macon, GA: Mercer University Press, 2004.

Frankenfield, Harry Crawford. *The Floods of the Spring of 1903, in the Mississippi Watershed.* United States Department of Agriculture, Washington, DC: Weather Bureau. 1904.

Fremling, Calvin R. *Immortal River: The Upper Mississippi in Ancient and Modern Times.* Madison: University of Wisconsin Press, 2005.

Gould, Emerson W. *Gould's History of River Navigation.* St. Louis, MO: Nixon-Jones Printing Co., 1889 (http://books.google.com/books?id=udyywXOVBvsC&pg=PA181&dq=1705+first+documented+mississippi+cargo&source=bl&ots=p5mLcWwnFk&sig=-Z_8oB)5ENBUANaimLUGX9drqpc&hl=en&ei=wBdvStquAojEMImN7OkI&sa=X&oi=book_result&ct=result&resnum=4).

Gunn, Angus M. *Unnatural Disasters: Case Studies of Human-Induced Environmental Catastrophes.* Westport, Conn.: Greenwood Press, 2003.

Hair, William Ivy. *The Kingfish and His Realm: The Life and Times of Huey P. Long.* Baton Rouge: Louisiana State University Press, 1996.

Harris, J. William. *Deep Souths: Delta, Piedmont, and Sea Island Society in the Age of Segregation.* Baltimore: John Hopkins University Press, 2003.

Hays, Samuel P. *Conservation and the Gospel of Efficiency: The Progressive Conservation Movement, 1890–1920.* Pittsburgh: University of Pittsburgh Press, 1999.

Hodge, Frederick Webb, ed. *Handbook of American Indians North of Mexico*, vol. 1. Washington DC: Government Printing Office, 1911.

Hudson, Charles M., and Carmen Chaves Tesser, eds. *The Forgotten Centuries:*

Indians and Europeans in the American South, 1521–1704. Athens: University of Georgia Press, 1994.

Hughes, Langston and E. B. Lewis (illustrator). *The Negro Speaks of Rivers.* New York: Hyperion, 2009.

Humphreys, A. A., and Henry Abbot. *Report Upon the Physics and Hydraulics of the Mississippi River; Upon the Protection of the Alluvial Region Against Overflow; and Upon the Deepening of the Mouths.* Washington, DC: Government Printing Office, 1867.

Hunter, Louis C., and Beatrice Jones Hunter. *Steamboats on the Western Rivers: An Economic and Technological History.* New York: Dover Publications, 1993.

Hurst, James Willard. *Law and Economic Growth: The Legal History of the Lumber Industry in Wisconsin, 1836–1915.* Cambridge, MA: Harvard University Press, 1964.

Jeansonne, Glen. *Leander Perez: Boss of the Delta.* Baton Rouge: Louisiana State University Press, 1977.

Kelman, Ari. *A River and Its City: The Nature of Landscape in New Orleans.* Berkeley: University of California Press, 2003.

Kern, William. *The Economics of Natural and Unnatural Disasters.* Kalamazoo, MI: W. E. Upjohn Institute, 2010.

Klein, Christine A., Federico Cheever, and Bret C. Birdsong. *Natural Resources Law: A Place-Based Book of Problems and Cases,* 2nd ed. New York: Aspen, 2009.

Landrum, Ney C. *The State Park Movement in America: A Critical Review.* Columbia: University of Missouri Press, 2004.

Lansden, John M. *A History of the City of Cairo, Illinois.* Carbondale: Southern Illinois University Press, 2009.

Lawson, Michael L. *Dammed Indians: The Pick-Sloan Plan and the Missouri River Sioux.* Norman: University of Oklahoma Press, 1982.

Leinwand, Gerald. *1927: High Tide of the Twenties.* New York: Four Walls Eight Windows, 2001.

Levitt, Jeremy I., and Matthew C. Whitaker. *Hurricane Katrina: America's Unnatural Disaster.* Lincoln, NE: University of Nebraska Press, 2009.

Lewis, Thomas A., *Brace for Impact: Surviving the Crash of the Industrial Age by Sustainable Living.* Parker, CO: Outskirts Press, 2009.

Louv, Richard. *Last Child in the Woods: Saving Our Children from Nature-Deficit Disorder.* Chapel Hill, NC: Algonquin Books, 2008.

Lewis, Peirce F. *New Orleans: The Making of an Urban Landscape.* Cambridge, MA: Ballinger, 1976.

Maney, Kevin. *Thomas Watson, Sr., and the Making of IBM: The Maverick and His Machine.* Hoboken, NJ: John Wiley & Sons, 2004.

McCook, Alistair. An Introduction to H. L. Mencken. *The Vintage Mencken.* New York: Vintage Books, 1955.

McGurty, Eileen. *Transforming Environmentalism: Warren County, PCBs, and the Origins of Environmental Justice*. Piscataway, NJ: Rutgers University Press, 2007.

McPhee, John. *The Control of Nature*. New York: Farrar, Straus, and Giroux, 1989.

National Research Council. *Mississippi River Water Quality and the Clean Water Act: Progress, Challenges, and Opportunities*. Washington, DC: National Academies Press, 2008.

——. *Missouri River Planning: Recognizing and Incorporating Sediment Management*. Washington, DC: National Academies Press, 2010,

O'Neill, Karen M. *Rivers by Design: State Power and the Origins of U.S. Flood Control*. Durham, NC: Duke University Press, 2006.

Otto, John Solomon. *The Final Frontiers, 1880–1930: Settling the Southern Bottomlands*. Westport, CT: Greenwood Press, 1999.

Petersen, William J. *Steamboating on the Upper Mississippi*. New York: Dover Publications, 1996.

Pisani, Donald. *Water and American Government: The Reclamation Bureau, National Water Policy, and the West*. Berkeley and Los Angeles: University of California Press, 2002.

Reed, Betsy, ed., *Unnatural Disaster: The Nation on Hurricane Katrina*. New York: Nation Books, 2006.

Reisner, Marc. *Cadillac Desert: The American West and Its Disappearing Water*. New York: Penguin Group (USA), 1993.

Sampson, R. Neil. *For Love of the Land: A History of the National Association of Conservation Districts*. League City, TX: Association, 1985.

Schneiders, Robert Kelley. *Unruly River: Two Centuries of Change along the Missouri*. Lawrence: University of Kansas Press, 1999.

Shindo, Charles J. *1927 and the Rise of Modern America*. Lawrence: University Press of Kansas, 2010.

Stegner, Wallace. *Beyond the Hundredth Meridian: John Wesley Powell and the Second Opening of the West*. New York: Houghton Mifflin, 1954.

Steinberg, Ted. *Acts of God: The Unnatural History of Natural Disaster in America*. New York: Oxford University Press, 2006.

Thorson, John E. *River of Promise, River of Peril: The Politics of Managing the Missouri River*. Lawrence: University Press of Kansas, 1994.

Tindall, George Brown. *The Emergence of the New South, 1913–1945*. Baton Rouge: Louisiana State University Press, 1967.

Twain, Mark. *Life on the Mississippi*. New York: Harper, 1951.

United Church of Christ, Commission for Racial Justice. *Toxic Waste and Race in the United States: A National Report on the Racial and Socio-Economic Characteristics of Communities with Hazardous Waste Sites*. New York: Public Data Access, Inquiries to the Commission, 1987.

U.S. Department of Agriculture (Weather Bureau). *The Floods of 1913 in the Rivers of the Ohio and Lower Mississippi Valleys.* Washington DC: Government Printing Office, 1913.

U.S. Environmental Protection Agency. *Environmental Equity: Reducing Risk for All Communities.* Washington, DC: US EPA, 1992.

Van Heerden, Ivor, and Mike Bryan. *The Storm: What Went Wrong and Why During Hurricane Katrina—The Inside Story from One Louisiana Scientist.* New York: Penguin, 2007.

Wark, James, and Joseph A. Mussulman. *Discovering Lewis and Clark from the Air.* Missoula, MT: Mountain Press Publishing, 2004.

Wright, Richard. *Uncle Tom's Children,* first Perennial ed. New York: HarperCollins, 2004.

World Bank, *Natural Hazards, Unnatural Disasters: The Economics of Effective Prevention.* Washington, DC: World Bank Publications, 2010.

Worster, Donald. *Dust Bowl: The Southern Plains in the 1930's.* New York: Oxford University Press, 1979.

Zakin, Susan, Bill McKibben, and Chris Jordan. *In Katrina's Wake: Portraits of Loss from an Unnatural Disaster.* Princeton, NJ: Princeton Architectural Press, 2006.

ARTICLES, REPORTS, AND WEBSITES

Abramovitz, Janet N., and Linda Starke. *Unnatural Disasters,* Worldwatch Paper 158, 2001.

Adams, Edward S., "At the End of Palsgraf, There is Chaos: An Assessment of Proximate Cause in Light of Chaos Theory." *University of Pittsburgh Law Review* 59 (1998).

Adamson, Erin. "Breaking Barriers: Topekans Reflect on Role in Desegregating Nation's Schools." *Topeka Capital Journal,* May 11, 2003.

American Society of Civil Engineers. "So You Live Behind a Levee! What You Should Know to Protect Your Home and Loved Ones from Floods," 2010 (http://content.asce.org/files/pdf/SoYouLiveBehindLevee.pdf).

America's Wetland Foundation. "About Us." (www.americaswetland.com/custompage.cfm?pageid=2).

———. "Louisiana Old River Control Complex and Mississippi River Flood Protection" (www.americaswetlandresources.com/background_facts/).

Annual Report of the Chief of Engineers, U.S. Army. *Mississippi River Commission.* Washington, DC: U.S. Government Printing Office, 1926.

Baker, Peter. "FEMA Director Replaced as Head of Relief Effort." *Washington Post,* September 10, 2005.

Barry, John M. "After the Deluge." *Smithsonian* 34 (2005).

———. "Battling Nature on the River." *Wall Street Journal,* April 30, 2011.

Barbier, Edward B., et al. "Economic Valuation of Wetlands: A Guide for Policy Makers and Planners," 1997 (http://www.ramsar.org/cda/en/

ramsar-pubs-books-economic-valuation-of-21378/main/ramsar/1-30-101%
5E21378_4000_0__).

Bearden, Russell E. "Arkansas' Worst Disaster: The Great Mississippi River Flood
of 1927." *Arkansas Review: A Journal of Delta Studies* 34 (August 2003).

Been, Vicki, "*Lucas v. The Green Machine*: Using the Takings Clause to Promote
More Efficient Regulation?" In *Property Stories*, eds. G. Korngold and A. Mor-
riss. New York: Foundation Press, 2004.

Begley, Sharon. "Man-Made Mistakes Increase Devastation of 'Natural' Disasters."
Wall Street Journal Online, September 2, 2005 (www.online.wsj.com/public/
article_print/SB112561128847329529.html).

Bell, Abraham, and Gideon Parchomovsky. "Givings." *Yale Law Journal* 111
(2001).

Bennett, Hugh. "Soil Erosion—A National Menace." *Scientific Monthly* 39
(November 1934).

Black, W. M. "The Problem of the Mississippi." *North American Review* 224
(1927).

"A Brief Chronology of What Congress Has Done Since 1824 to Control the
Floods of the Mississippi." *Congressional Digest* 7 (1928).

Brown, D. Clayton. "Western Tributaries of the Mississippi." *National Waterways
Study*. U.S. Army Corps of Engineers, 1983 (http://www.iwr.usace.army.mil/
docs/iwrreports/WESTERNTRIBUTARIESOFTHEMISSISSIPPIJANUARY
1983.pdf).

Brown, Robert M. "The Mississippi River Flood of 1912." *Bulletin of the American
Geographical Society* 44.9 (1912).

Bullard, Robert D., and Beverly Wright. "Disastrous Response to Natural and
Man-Made Disasters: An Environmental Justice Analysis Twenty-Five Years
after Warren County." *UCLA Journal of Environmental Law and Policy*, 26
(2008).

Burgess, Barbara. "Topeka's Roots: The Prairie Potato" (2003) (www.barbburgess
.com/research-topics/prairie-potato-topeka/topeka-s-roots-the-prairie-potato).

Byrne, J. Peter, "Ten Arguments for the Abolition of the Regulatory Takings Doctrine." *Ecology L.aw Quarterly* 22 (1995).

Caffey, Rex. "Louisiana Hurricane Resources, Barrier Islands & Wetlands." Louisiana Sea Grant College Program, LSU Ag Center , 2005 (www.laseagrant.org/
hurricane/archive/wetlands.htm).

Cardi, Jonathan. "Reconstructing Foreseeability." *Boston College Law Review* 46
(2004–2005).

Carrns, Ann. "Holes in the Dike: Long before Flood, New Orleans was Prime for
Leaks." *Wall Street Journal,* November 25, 2005.

Chavis, Benjamin. "Concerning the Historical Evolution of the 'Environmental
Justice Movement' and the Definition of the Term: 'Environmental Racism'"
(www.drbenjaminchavis.com/pages/landing/?blockID-73318&feedID=3359).

Cheshes, Jay. "Annie Gunn's: Chesterfield, Mo." *Nation's Restaurant News*, May 22, 2006.

Cherny, Robert W. "Graft and Oil: How Teapot Dome Became the Greatest Political Scandal of Its Time." *History Now*. Gilder Lehrman Institute of American History, 2009.

Chin, Gabriel J. "The Jena Six and the History of Racially Compromised Justice in Louisiana." *Harvard Civil Rights-Civil Liberties Law Review* 44 (2009).

"Citizens Fight Flood Along with Convicts; Rich and Poor at Baton Rouge Unite in Desperate Struggle." *New York Times*, May 4, 1912.

City of Chesterfield. "History" (www.chesterfield.mo.us/history.html).

City of Hardin. "Hardin Cemetery History." (www.cityofhardin.com/cemeteries/hardin-cemetery/history.html).

Coastal Protection and Restoration Authority of Louisiana. "Louisiana Coastal Facts" (http://dnr.louisiana.gov/assets/OCM/OCM/webfactsheet_20110727.pdf).

Colten, Craig E. "Environmental Justice on the American Bottom." In *Common Fields: An Environmental History of St. Louis,* eds. Robert Archibald, et al. St. Louis: Missouri Historical Society Press, 1997.

Combs, Lanna, and Charles Perry. "The 1903 and 1993 Floods in Kansas—The Effects of Changing Times and Technology." *U.S. Geological Survey, Fact Sheet 019-0* (March 2003).

"Completion of the Jetties." *New York Times Tribune*, July 15, 1897.

"Comprehensive and Integrated Approach to Meet the Water Resources Needs in the Wake of Hurricanes Katrina and Rita." Statement of Scott Faber. Hearings before the Senate Committee on Environmental and Public Works, 109th Congress, 2005 (http://epw.senate.gov/hearing_statements.cfm?id=248649).

Congressional Research Service, Report for Congress. "Federal Flood Insurance: The Repetitive Loss Problem," 2005 (www.fas.org/sgp/crs/misc/RL32972.pdf).

Costanza, Robert, et al., "The Value of the World's Ecosystem Services and Natural Capital." *Nature*, May 15, 1997.

Craig, Robin Kundis. "Public Trust and Public Necessity Defenses to Takings Liability for Sea Level Rise Responses on the Gulf Coast." *Journal of Land Use and Environmental Law* 26 (2011).

Curley, John. "Annie Gunn's Big Bounce Back." *St. Louis Post-Dispatch*, September 20, 2000.

Dahl, Thomas E. "Status and Trends of Wetlands in the Conterminous United States 1998 to 2004." U.S. Fish & Wildlife Service, 2006.

Dahl, Thomas E., and Gregory J. Allord. "History of Wetlands in the Conterminous United States." U.S. Geological Survey Water Supply Paper 2425, U.S. Fish & Wildlife Service, March 7, 1997.

Deslodge, Rick. "St. Louis Character: Thom Sehnert." *St. Louis Business Journal*, August 17, 2005.

"Denies 300 Drowned in Southern Flood; Gov. Brewer of Mississippi Telegraphs The Times That Report Is Untrue." *New York Times*, April 21, 1912.

Department of Health and Human Services, "A Chronology of Major Events Affecting the National Flood Insurance Program," 2002 (www.dhs.gov/xlibrary/assets/privacy/privacy_pia_mip_apnd_h.pdf).

Department of the Interior, National Park Service. "Finding of No Significant Impact, Potomac Park Levee System Improvements." May 4, 2009 (http://www.ncpc.gov/DocumentDepot/PublicNotices/Potomac_Park_Levee_FONSI_6885_May2009.pdf).

Diamond, Jared. "The Last Americans, Environmental Collapse and the End of Civilization." *Harper's Magazine*, June 2003.

Driesen, David M., et al. "An Unnatural Disaster: The Aftermath of Hurricane Katrina." Center for Progressive Reform, 2005 (www.progressivereform.org/articles/unnatural_Disaster_512.pdf).

Earth Economics. "Report on the Mississippi River Delta" (www.eartheconomics.org/FileLibrary/file/Reports/Louisiana/Earth_Economics_Report_on_the_Mississippi_River_Delta_compressed.pdf).

"Farm Hands Perish at Scott Crevasse: Furious Waters of Mississippi Trap Negro Laborers." *Memphis Commercial Appeal*, April 22, 1927.

Federal Emergency Management Agency. "Report on Flooding and Stormwater in Washington, D.C.," 2008 (www.ncpc.gov/DocumentDepot/Publications/FloodReport2008.pdf).

Federal Reserve Bank of Minneapolis. "What Is a Dollar Worth?" April 4, 2011 (http://minneapolisfed.org/research/data/us/calc/).

Finkel, Jori. "Is Everything Sacred? A Respected Art Dealer is Busted for Selling a Cheyenne War Bonnet." *Legal Affairs* (July/August 2003).

Fischer, Claude S. "Technology's Retreat: The Decline of Rural Telephony in the United States, 1920–1940." *Social Science History* 11 (Autumn 1987).

Fitch, George. "The Missouri River: Its Habits and Eccentricities Described by a Personal Friend." *American Magazine* 53.6 (1907).

Fletcher, Michael A., and Richard Morin. "Bush's Approval Rating Drops to New Low in Wake of Storm." *Washington Post*, September 13, 2005.

"Flood Control in the Mississippi Valley." House of Representatives Doc. No. 90, 70th Cong., 1st Sess., December 1, 1927.

"Flood Death Toll Numbers Hundreds; Great Torras Crevasse Alone Takes Scores of Lives in Louisiana Lowlands." *New York Times*, May 7, 1912.

"Flood Near Crest; Hunger Kills Scores." *New York Times*, May 8, 1912.

Fothergill, A., E. G. M. Maestas, and J. D. Darlington. "Race, Ethnicity, and Disasters in the United States: A Review of the Literature." *Disasters* 23.2 (1999).

"Fresh Perils Added to Flood by Storm; Severest Rain in History of New Orleans Renews Danger in Inundated Southern Territory." *New York Times*, May 11, 1912.

Futrell, J. William. "The IUCN Sustainable Soil Project and Enforcement Failures." *Pace Environmental Law Review* 24 (2007).

Galloway, Gerald E., Jr. "Corps of Engineers Responses to the Changing National Approach to Floodplain Management since the 1993 Midwest Flood." *Journal of Contemporary Water Resources and Education* (March 2005).

General Accounting Office. "Army Corps of Engineers: Lake Pontchartrain and Vicinity Hurricane Protection Project." Testimony of Anu Mittal Before the Subcommittee on Energy and Water Development, Committee on Appropriations, House of Representatives, GAO publication 05-1050T, September 28, 2005 (http://www.gao.gov/new.items/d051050t.pdf).

Germano, Nancy M. "White River and the 'Great Flood' of 1913: An Environmental History of Indianapolis," 2010 (www.mchsindy.org/history/1913flooding_article.pdf).

Gresko, Jessica. "Work to Begin on National Mall Levee." *Associated Press*, November 15, 2010.

Goodwin, Derek V. "Raiders of the Sacred Sites." *New York Times Magazine*, December 7, 1986 (www.nytimes.com/1986/12/07/magazin/raiders-of-the-sacred-sites.html?scp=3&sq=derek+%22raiders+of+the+sacred%22&st=nyt).

Gordon, Larry. "Top Court to Hear Church Camp Plea." *Los Angeles Times*, July 10, 1986.

Gorton, D. "Methods and Materials [of J. C. Coovert]." Lecture, May 22, 2003 (www.jccoovert.com/lecture/lecture08.html).

Grubaugh, J. W., and R. V. Anderson. "Long Term Effects of Navigation Dams on a Segment of the Upper Mississippi River." *Regulated Rivers: Research and Management* 4 (1989).

Grunwald, Michael. "Washed Away." *New Republic*, October 27, 2003.

Guarino, Mark. "Mississippi River Flooding: After Levee Blast, Threat Shifts to Memphis." *Christian Science Monitor*, May 3, 2011.

Halper, Louise A. "Why the Nuisance Knot Can't Undo the Takings Muddle." *Indiana Law Review* 28 (1995).

Hanson Professional Services, Inc. "Monarch-Chesterfield Levee Flood-Wall Design" (www.hanson-inc.com/projects.aspx?projectid=06s3035a).

"Harlem, MO Flood, May 1903," July 26, 2008 (www3.gendisasters.com/Missouri/8083/harlem-mo-flood-may-1903).

Hinrichsen, Don. "The Coastal Population Explosion." *Trends and Future Challenges for U.S. National Ocean and Coastal Policy* (1999).

Historic American Engineering Record, Lake Winnibigoshish Reservoir Dam, Deer River Vicinity, Itasca County, MN, Survey No. HAER MN-65, 1968 (http://hdl.loc.gov/loc.pnp/hhh.mn0389).

Hodges, Leah. "New Orleans Evacuees and Activists Testify at Explosive House Hearing on the Role of Race and Class in Government's Response

to Hurricane Katrina." *Democracy Now!* December 9, 2005 (http://www
.democracynow.org/article.pl?sid=05/12/09/1443240).

Houck, Oliver. "Floodway into the Atchafalaya Basin Saves New Orleans," May
6, 2011 (www.nola.com/opinions/index.ssf/2011/05/floodway_into_the_
atchafalaya.html).

———. "Rising Water: The National Flood Insurance Program and Louisiana."
Tulane Law Review 60 (1985).

Hsu, Spencer S., and Susan B. Glasser. "FEMA Director Brown Singled Out by
Response Critics." *Washington Post*, September, 6, 2005.

"Human Dike Used to Hold Back Flood." *New York Times*, April 11, 1912 (http://
query.nytimes.com/mem/archive-free/pdf?res=9E0CEFD6143CE633A25752
C1A9629C946396D6CF).

"Hundreds in Trap as Levee is Blown Away; Army Men Use 2,000 Pounds of
Dynamite in Effort to Save Cairo, Illinois." *New York Times*, January 26, 1937.

Independent Levee Investigation Team. "New Orleans Levee Systems: Hurricane
Katrina," July 31, 2006 (www.ce.berkeley.edu/~new_orleans/report/CH_4.pdf).

Interagency Floodplain Management Review Committee. "Sharing the Challenge:
Floodplain Management into the 21st Century." Gallup Report, June 1994.

International Communist League. "Black Oppression and the Great Mississippi
Flood of 1927." *Workers Vanguard*, April 14, 2006 (www.icl-fi.org/english/wv/
868/1927.html).

"Island 37." *The Encyclopedia of Arkansas History & Culture*, 2008 (http://
encyclopediaofarkansas.net).

Jarrell, Jerry D., Max Mayfield, and Edward N. Rappaport. "The Deadliest, Costliest, and Most Intense United States Hurricanes From 1900 to 2000." NOAA
Technical Memorandum NWS TPC-1, October 2001.

John Handcox, "The Planter and the Sharecropper." Labor Notes (www.labor
notes.org/node/893).

Johnson, Gary P. Robert R. Holmes, Jr., and Lloyd A. Waite. "The Great Flood of
1993 on the Upper Mississippi River—10 Years Later." USGS Fact Sheet 2004-
3024, May 2004.

Jones, Terry L. "Looming Menace?" *Reel Louisiana Adventures*, August 2, 2010
(http://69.16.203.116/details.php?id=2430)

Josephson, D. H. "The Great Midwest Flood of 1993, Natural Disaster Survey
Report." Department of Commerce, NOAA-National Weather Service (1994).

"Katrina's Unlearned Lessons." *Washington Post*, June 7, 2006.

Kidder, Tristram R. "Making the City Inevitable: Native Americans and the
Geography of New Orleans." In *Transforming New Orleans and Its Environs:
Centuries of Change,* ed. Craig E. Colten. Pittsburgh: University of Pittsburgh
Press, 2000.

King, Rawle O. "Federal Flood Insurance: The Repetitive Loss Problem." Congressional Research Service, RL32972, June 30, 2005.

——. "National Flood Insurance Program: Background, Challenges, and Financial Status." Congressional Research Service, Report 7-5700, R40650, July 1, 2011.

Kingsolver, Barbara. "Fresh Water." *National Geographic* 46 (April 2010).

Klein, Christine A. "On Dams and Democracy." *Oregon Law Review* 78 (1999).

Knabb, Richard D., Jamie R. Rhome, and Daniel P. Brown. "Tropical Cyclone Report: Hurricane Katrina, National Hurricane Center," December 20, 2005, and August 10, 2006 update.

Krier, James E. "The Takings-Puzzle Puzzle." *William & Mary Law Review* 38 (1997).

Krupa, Michelle. "Ethel Williams Dies at 75: 9th Ward Resident Took Bush to Flooded Home." (New Orleans) *Times-Picayune*, January 28, 2009.

Kunreuther, Howard, "Has the Time Come for Comprehensive Natural Disaster Insurance?" In *On Risk and Disaster: Lessons from Hurricane Katrina,* eds. Ronald J. Daniels, Donald F. Ketti, Howard C. Kunreuther, and Amy Gutmann. Philadelphia: University of Pennsylvania Press, 2006.

Louisiana Coastal Wetlands Planning Protection & Restoration Act Program, "Mississippi River Basin Dynamics" (http://lacoast.gov/new/about/basin_data/mr/default.aspx).

——. "Mississippi River Delta Basin" (www.lacoast.gov/landchange/basins/mr/).

Lawson, Michael L. "Federal Water Projects and Indian Lands: The Pick-Sloan Plan, A Case Study." *American Indian Culture and Research Journal* 7.1 (1983).

Leonard, Mike. "1913 Flood: Devastating, but Bloomington Caught a Break." *Bloomington Herald Times*, June 12, 2008.

Leopold, Aldo. "The Conservation Ethic." *Journal of Forestry* 31 (October 1933).

Libecap, Gary D., and Zeynep Kocabiyik Hansen. " 'Rain Follows the Plow' and Dryfarming Doctrine: The Climate Information Problem and Homestead Failure in the Upper Great Plains, 1890–1925." *Journal of Economic History* 62 (2002).

Library of Virginia. "The Virginia Newspaper Project Examines the News Covering the Sinking of R.M.S. *Titanic*, April 14, 1912" (http://www.lva.virginia.gov/exhibits/titanic/titanic1.htm).

Louisiana Department of Natural Resources, Coastal Protection and Restoration Authority. "Louisiana Coastal Facts" (http://dnr.louisiana.gov/assets/OCM/OCM/webfactsheet_20110727.pdf).

Maantay, Juliana, and Andrew Maroko. "Mapping Urban Risk: Flood Hazards, Race, and Environmental Justice in New York." *Applied Geography* 29.1 (January 1, 2009).

Mank, Bradford. "Title VI and the Warren County Protests." *Golden Gate University Environmental Law Journal* 1.1 (2007).

Marshall, Bob. "City's Fate Sealed in Hours: Timeline Maps Course of Post-Katrina Deluge." (New Orleans) *Times-Picayune*, May 14, 2006.

Marshall, Edward. "Willcocks Sees Simple Solution of Mississippi Flood Problem; Sir William Willcocks, Designer of the Assouan Dam, Says Our Problem Is Easier Than That of the Mesopotamian Rivers, Solved 5,000 Years Ago." *New York Times Magazine*, May 24, 1914 (http://query.nytimes.com/gst/abstract .html?res=9F01EFDF1F39E633A25757C2A9639C946596D6CF&scp=7&sq =1912+flood+levee&st=p).

Martinsdale, Bryan, and Paul Osman. Illinois Association for Floodplain and Stormwater "Why the Concerns with Levees? They're Safe, Right?" *IAFSM Newsletter*, Fall 2007 (www.illinoisfloods.org/documents/IAFSM_Levee%20 Article.pdf).

McGurty, Eileen Maura. "From NIMBY to Civil Rights: The Origins of the Environmental Justice Movement." *Environmental History* 2.3 (1997).

McQuaid, John, and Mark Schleifstein. "In Harm's Way: Surging Water Is the Biggest Threat to New Orleans." (New Orleans) *Times-Picayune*, June 23, 2002.

———. "Left Behind." (New Orleans) *Times-Picayune*, June 24, 2002.

McGee, W. J. "The Flood Plains of Rivers." *Forum* 11 (1891).

"Memphis in Grave Peril; Levee Near Wilson, Ark., Goes Out—Other Levees Menaced." *New York Times*, April 9, 1913.

Mercer, David. "Cairo Evacuated: Mayor Tells Residents to Leave Flood-Threatened Illinois Town," May 1, 2011 (http://www.huffingtonpost.com/ 2011/05/01/cairo-evacuated-mayor-tel_n_855988.html).

Michel, Mark. "The Archaeological Conservancy and Site Protection." In *Protecting the Past,* eds. George S. Smith and John E. Ehrenhard. St. Louis: Missouri Historical Society Press, 1997 (www.nps.gov/history/seac/protecting/html/ 5j-michel.htm).

Minnesota Historical Society. "Mill City Museum" (http://events.mnhs.org/media/ Kits/Sites/mcm/background.htm).

"Mississippi Rising: Apocalypse Now?" April 28, 2011 (http://www.dailyimpact .net/2011/04/28/mississippi-rising-apocalypse-now/).

Mississippi River Commission. "The Mississippi River & Tributaries Project: Birds Point-New Madrid Floodway," Information Paper (http://www.mvd .usace.army.mil/mrc/mrt/Docs/Birds%20Point-New%20Madrid%20info%20 paper%20FINAL%200426.pdf).

———. "The Mississippi River & Tributaries Project: Floodways." Information Paper, October 2007 (www.mvd.usace.army.mil/mrc/mrt/Docs/Floodways%20 info%20paper.pdf).

"Missouri River Mainstem Reservoir System Master Water Control Manual," March 19, 2004 (www.nwd-mr.usace.army.mil/rcc/reports/MManual/Master% 20Manual.pdf).

Missouri State Emergency Management Agency. "10-Year Anniversary of the '93– '94 Floods" (www.sema.dps.mo.gov/flood%20anniversary.pdf).

———."Missouri Hazard Analysis." Annex B: Riverine Flooding B-5, 2006 (www.sema.dps.mo.gov/HazardAnalysis/AnnexB.pdf).

Morris, Christopher. "Impenetrable but Easy." In *Transforming New Orleans and Its Environs: Centuries of Change*, ed. Craig E. Colten. Pittsburgh: University of Pittsburgh Press, 2000.

Nash, Steve. "Water World: The Coming Seawall Craze." *New Republic*, September 24, 2010.

National Aeronautic and Space Administration, Soil Science Education. "In the Time It Took to Form One Inch of Soil," 2009 (http://soil.gsfc.nasa.gov/index .php?section=74).

National Commission on the BP Deepwater Horizon Spill and Offshore Drilling. "Final Report," January 11, 2011.

National Hurricane Center. "Hurricane History: Katrina 2005" (www.nhc.noaa .gov/HAW2/english/history.shtml#katrina).

National Museum of American History, Behring Center, Topeka Kansas. "Segregation in the Heartland" (americanhistory.si.edu/brown/history/4-five/topeka -kansas-2.html).

National Oceanic and Atmospheric Administration. "Historic Flood Events in the Missouri River Basin" (www.crh.noaa.gov/mbrfc/?n=flood).

———."National Overview in Population Trends," 2005 (oceanservice.noaa.gov/ programs/mb/pdfs/2_national_overview.pdf).

———."Population Trends along the Coastal United States: 1980–2008" (http:// oceanservice.noaa.gov/facts/population.html).

———."State of the Coast: U.S Population Living in Coastal Watershed Counties" (http://stateofthecoast.noaa.gov/population/welcome.html).

———."United States Flood Loss Report—Water Year 2011" (http://pubs.usgs .gov/pp/1798c/pp1798c.pdf).

———.National Weather Service Weather Forecast Office, Louisville, Kentucky, "The Great Flood of 1937" (www.crh.noaa.gov/lmk/?n=flood_37).

National Park Service. "Explore Geology" (http://www.nature.nps.gov/geology/ parks/icag/index.cfm).

———."Ice Age, National Scenic Trail" (http://nps.gov/iatr/index.htm).

National Wildlife Federation. "NWF Plays Key Role in Program to Move Buildings out of Flood Plains," *National Wildlife,* February–March1997 (www.nwf.org/ nationalwildlife/article.cfm?issueID=53&articleID=659#key).

The Nature Conservancy. "The Mississippi River and its Floodplain: Restoring Connections for People and Nature" (http://www.nature.org/ourinitiatives/ habitats/riverslakes/explore/mississippi-river-and-its-floodplain-restoring- connections-for-people-and.xml).

———."The Upper Mississippi River Program" (http://www.nature.org/ ourinitiatives/regions/northamerica/unitedstates/iowa/placesweprotect/upper -mississippi-river.xml).

Negri, A., et al. "The Hurricane-Flood-Landslide Continuum." *Bulletin of the American Meteorological Society* 86.9 (2005).

Neibauer, Michael. "Planners Settle on Two Designs for the New National Mall Levee." *Washington Examiner*, December 4, 2008.

NET Nebraska. "1952 Missouri River Flood: A Look Back," April 5, 2002 (http://netnebraska.org/interactive-multimedia/news/net-news-statewide-1952 -missouri-river-flood-look-back-4502?field_media_categories_tid_1=All&title= &tid=&page=8).

"New Orleans' Fate Depends on a Train." *New York Times*, April 10, 1913.

"New Orleans Raises Dikes; Expects to Have Them Well above Mississippi's Flood Crest." *New York Times*, April 12, 1913.

Norfleet, Nicole. "Melting Snow Creates Anxiety at C&O Canal's Potomac Gate." *Washington Post*, March 16, 2010.

North Carolina Division of Waste Management. "Warren County PCB Landfill Fact Sheet" (http://wastenot.enr.state.nc.us/WarrenCo_Fact_Sheet.htm).

North Carolina State University, Water Quality Group. "Wetland Functions (or Processes) and Values" (http://www.water.ncsu.edu/watershedss/info/wetlands/ funval.html).

Novitzki, Richard P., et al. "Restoration, Creation, and Recovery of Wetlands: Wetland Functions, Values, and Assessment." U.S. Geological Survey Water Supply Paper 2425 (http://water.usgs.gov/nwsum/WSP2425/functions.html).

Nulik, Jeremy. "Chesterfield Valley Awakening Continues." *St. Louis Small Business Monthly,* May 2009.

Ohio Historical Society. "Severe Weather in Ohio" (www.ohiohistory.org/etcetera/ exhibits/swio/pages/content/1913_flood.htm).

———. "1913: Statewide Flood" (www.ohiohistory.org/etcetera/exhibits/swio/ pages/content/1913_flood.htm).

O'Neil, Sandra George. "Superfund: Evaluating the Impact of Executive Order 12898." *Environmental Health* 115 (2007).

Pearcy, Matthew T. "A History of the Ransdell-Humphreys Flood Control Act of 1917." *Louisiana History: The Journal of the Louisiana Historical Association* 41.2 (Spring 2000).

Paschall, Gabrielle. "Protecting Our Past: The Need for Uniform Regulation to Protect Archaeological Resources." *Thomas M. Cooley Law Review* 27 (2010).

"Peonage Bill Hits Workers in Louisiana: Plantation Laborers Are Made Serfs." *Chicago Defender*, August 21, 1926.

Perry, Charles A. "Significant Floods in the United States during the 20th Century: USGS Measures a Century of Floods." USGS Fact Sheet 024-00, March 2000.

PETE: National Partnership for Environmental Technology Education. "NSF Tribal" (www.nationalpete.org/index.php?option=com_content&view=article& id=84&Itemid=87).

Philippi, Nancy. "Plugging the Gaps in Flood-Control Policy." *Issues in Science and Technology* 11.2 (December 1, 1994).

Pinter, Nicholas, et al. "Flood Trends and River Engineering on the Mississippi River System." *Geophysical Research Letters* 35 (2008).

Pitlick, John. "A Regional Perspective of the Hydrology of the 1993 Mississippi River Basin Floods." *Annals of the Association' of American Geographers* 87 (1997).

Pitluk, Adam. "Dammed to Eternity." *Riverfront Times*, January 26, 2000.

——. "Revisiting the Great Flood of 1993 and James Scott." *Huffington Post*, June 12, 2007 (www.huffingtonpost.com/adam-pitluk/revisiting-the-great-floo_b_51842.html).

——. "Scapegoat, James Scott Will Never Forget the Great Flood of '93, He Has No Choice." *Illinois Times*, January 29, 2006.

"Planners Settle on Two Designs for the New National Mall Levee." *Washington Examiner*, December 4, 2008.

Platt, Rutherford H., "Review of Sharing the Challenge: Floodplain Management into the 21st Century." *Environment* 37.25 (1995).

Posner, Richard A. "A Theory of Negligence." *Journal of Legal Studies* 1 (1972).

Power, Stephen, John Kell, and Siobhan Hughes. "BP, Oil Industry Take Fire at Hearing," *Wall Street Journal*, June 16, 2010.

"President Coolidge's Analysis of the Mississippi Flood Control Problem, as Contained in His Annual Message to Congress, Dec. 6, 1927." *Congressional Digest* 7 (1928).

Reuss, Martin. "Andrew A. Humphreys and the Development of Hydraulic Engineering: Politics and Technology in the Army Corps of Engineers, 1850–1950." *Technology and Culture*, 26 (1985).

Roback, Jennifer. "Southern Labor Law in the Jim Crow Era: Exploitative or Competitive?" *University of Chicago Law Review* 51 (1984).

Robarge, Drew. "Washington D.C.'s 19th Century Reclamation Project, Making Room for Blossoms and Monuments." *Atlantic*, March 28, 2011. (www.theatlantic.com/technology/archive/2011/03/washington-dcs-19th-century-reclamation-project/73078/).

Roll, Jarod. "Out Yonder on the Road": Working Class Self-Representation and the 1939 Roadside Demonstration in Southeast Missouri." *Southern Spaces*, March 16, 2010.

Rose, Carol M. "Mahon Reconstructed: Why the Takings Doctrine Is Still a Muddle." *Southern California Law Review* 57 (1984).

Sagarra, Susan E. "Louis Sachs: The Man with a Plan." *West Newsmagazine*, February 15, 2009.

Salzman, James. "Thirst: A Short History of Drinking Water." *Yale Journal of Law and Humanities* 18 (2006).

Sanders, J. Y. "Gov. Sanders Tells of Big Flood Losses; 375,000 Persons Homeless and Property Damage $6,000,000." *New York Times*, May 8, 1912.

Sanders, Robert. "UC Berkeley-Led Levee Investigation Team Releases Final Report at Public Meeting in New Orleans." Press release, May 20, 2006 (www.berkeley.edu/news/media/releases/2006/05/24_leveereport.shtml).

Scales, Adam F. "A Nation of Policyholders: Governmental and Market Failure in Flood Insurance." *Mississippi Central Law Review* 26 (2006–2007).

Schleifstein, Mark. "Corps Report Ignores Call for Specifics; Details of Category 5 Protections Left Out." (New Orleans) *Times-Picayune*, July 1, 2006.

———, and Keith Darce. "Goodbye Mister Go: Economics, Environment May Doom Shipping Channel, New Orleans." (New Orleans) *Times-Picayune*, October 18, 1998.

Schor, Elana. "Superfund: At Midlife, Unending Cleanups and Less 'Real Money.'" *Greenwire*, December 10, 2010 (http://www.eenews.net/public/Greenwire/2010/12/10/1?page_type=print).

Schwartz, John. "Civil Lawsuit Over Katrina Begins." *New York Times*, April 21, 2009.

Sea Grant. "Urban Coasts" (http://www.seagrant.noaa.gov/other/admininfo/historical/theme_teams/urban%20coasts.pdf).

Seed, R. B., et al. "Independent Levee Investigation Team Final Report: Investigation of the Performance of the New Orleans Flood Protection System in Hurricane Katrina on August 29, 2005," 2006 (http://www.ce.berkeley.edu/projects/neworleans/).

Sendzimir, Jan, Steven Light, and Karolina Szymanowska. "Adaptive Understanding and Management for Floods," 1999 (www.adaptivemanagement.net/Flooding.doc).

Shallat, Todd. "In the Wake of Hurricane Betsy." In *Transforming New Orleans and Its Environs: Centuries of Change*, ed. Craig E. Colten. Pittsburgh: University of Pittsburgh Press, 2000.

Sioux City History, "Disasters" (http://www.siouxcityhistory.org/disasters).

Singer, Saul Jay. "Flooding the Fifth Amendment: The National Flood Insurance Program and the Takings Clause." *Boston College Environmental Affairs Law Review* 17 (1990).

Smith, Aaron Lake. "Trying to Revitalize a Dying Small Town." *Time Magazine*, February 15, 2010.

South Carolina State Climatology Office, Hurricanes. "South Carolina Hurricane Climatology-Notable South Carolina Hurricanes" (http://www.dnr.sc.gov/climate/sco/Tropics/hurricanes_affecting_sc.php).

Steinzor. Rena I. "Unfunded Environmental Mandates and the New (New) Federalism: Devolution, Revolution, or Reform?" *Minnesota Law Review* 81 (1996).

Sulzberger, A. G. "Army Corps Blows Up Missouri Levee." *New York Times*, May 2, 2011.

Talmadge, Alice. "Dust, Drought and Despair." *Forest Magazine*, Winter 2008 (www.fseee.org/forest-magazine/200249).

Tarlock, A. Dan. "A First Look at a Modern Legal Regime for a 'Post-Modern' United States Army Corps of Engineers." *Kansas Law Review* 52 (2004).

———. "The Missouri River: The Paradox of Conflict Without Scarcity." *Great Plains Natural Resources Journal* 2 (Spring 1997).

Taylor, Harry. "Problems of Flood Control." *Scientific Monthly* 16.4 (April 1923).

"Testimony of Mr. Maxwell before the Senate Commerce Committee on an Act to Provide for the Control of the Floods of the Mississippi River and of the Sacramento River, California, and for Other Purposes." *Senate Documents, 64th Congress, 1st Session* 43 (December 19, 1916).

THF Realty. "Chesterfield Commons," (developer's website) (www.thfrealty.com/index.php/our-properties/property/chesterfield-commons/).

Thorson, Norman W. "Damned If You Do, Damned If You Don't—Reflections on John Ferrell's Big Dam Era." *Great Plains Natural Resources Journal* 2 (1997).

Tibbetts, John. "Waterproofing the Midwest." *Planning*, April 1994.

Townsend, Frances Fragos. "The Federal Response to Hurricane Katrina: Lessons Learned," 2006 (library.stmarytx.edu/acadlib/edocs/katrinawh.pdf).

UNC Exchange Project. "Real People—Real Stories: Seeking Environmental Justice, Afton, NC." Warren County, 2006 (www.exchangeproject.unc.edu/documents/pdf/real-people/Afton%20summary%2007-0320%20for%20web.pdf).

University of Wisconsin-Stevens Point. "Oxbow Lake Formation," 2011 (www.uwsp.edu/geo/faculty/lemke/geog101).

Upin, Bruce. "A River of Subsidies." *Forbes*, March 23, 1998.

U.S. Army Corps of Engineers. "Federal Participation in Waterways Development," 1985 (www.mvn.usace.army.mil/PAO/history/MISSRNAV/federal.asp).

———. "Global Security," updated May 7, 2011 (http://www.globalsecurity.org/military/agency/army/usace.htm).

———. "Information About MRGO" (www.mrgo.usace.army.mil/default.aspx?p+MRGOInfo).

———. "Lake Pontchartrain and Vicinity Hurricane Protection, Orleans Parish" (www.mvn.usace.army.mil/pd/projectslist/home.asp?projectID=217).

———. "Louisiana Coastal Area Ecosystem Restoration Study" (www.mvn.usace.army.mil/environmental/lca.asp).

———. "The Mississippi Drainage Basin" (www.mvn.usace.army.mil/bcarre/missdrainage.asp).

———. "Mississippi River Navigation," 2008 (www.mvn.usace.army.mil/pao/history/MISSRNAV/index.asp).

———. "Missouri River Mainstem Reservoir System Master Water Control Manual," March 19, 2004 (www.nwd-mr.usace.army.mil/rcc/reports/MManual/Master%20Manual.pdf).

U.S. Army Corps of Engineers. "Old River Control" (www.mvn.usace.army.mil/
recreation/rec_oldrivercontrol.asp).

———. "Performance Evaluation of the New Orleans and Southeast Louisiana Hurricane Protection System: Final Report of the Interagency Performance Evaluation Task Force." I-3-4 (2006).

———. "News Release, Army Corps of Engineers, Baltimore District, Awards $4.6 Million Contract to Help Protect National Mall from River Flooding." September 17, 2010 (www.nab.usace.army.mil/Publications/News/10/10-15.pdf).

———, Headquarters. "Mission & Vision" (http://www.usace.army.mil/about/
missionandvision.aspx).

———, Interagency Performance Evaluation Task Force. "Repair Improvements for the New Orleans Hurricane Protection System." News Release No. PA-05-16, December 9, 2005.

———, Mississippi Valley Division. "Chesterfield, Missouri" (www.mvs.usace.army
.mil/pm/project%20menu/chesterfield/Mainframe.htm).

———. "Introduction." (http://www.mvd.usace.army.mil/welcome.htm).

———. "Welcome." (http://www.mvd.usace.army.mil/welcome.htm).

———, New Orleans District. "The Mississippi Drainage Basin" (http://www.mvn
.usace.army.mil/bcarre/missdrainage.asp).

———. "Mississippi Flood Control: The Mississippi River" (http://www.mvn.usace
.army.mil/bcarre/missriver.asp).

———. "The Mississippi River" (http://www.mvn.usace.army.mil/bcarre/missriver
.asp).

———. "The Mississippi River and Tributaries (MR&T) Project," updated August 25, 2011 (http://www.mvn.usace.army.mil/bcarre/missproj.asp).

———. "The Mississippi River and Tributaries Project," May 19, 2004 (http://www
.mvn.usace.army.mil/pao/bro/misstrib.htm).

———, Rock Island District. "Levee Safety Program: Levee Terms & Definitions" (www.mvr.usace.army.mil/publicaffairsoffice/LSP1/LSPLeveeTerms.htm).

———, St. Louis District. "Chesterfield" (www.mvs.usace.army.mil/pm/project%20
menu/chesterfield/Mainframe.htm).

———. "Monarch-Chesterfield Levee, Missouri" (www.mvs.usace.army.mil/pm/
monarch-chesterfield/index.html).

———, St. Paul District. "Brief History," 2008 (www.mvp.usace.army.mil/history).

———. "Mississippi Locks and Dams" (www.mvp.usace.army.mil/navigation/
default.asp?pageid=145).

———. "Regulating Mississippi River Navigation Pools" (http://www.mvp-wc.usace
.army.mil/projects/general/ld_brochure.html).

———, Team New Orleans. "Bonnet Carré Spillway Overview: Site Selection, Design Advances & Project Statistics," August 29, 2011 (/www.mvn.usace.army
.mil/bcarre/designadvances.asp).

——. "Spillway Operation Information," August 29, 2011 (http://www.mvn.usace
.army.mil/bcarre/spilloperation.asp).

——, Vicksburg District. "A Century of Service" (www.mvk.usace.army.mil/index
.php?pID=wwa&s=4).

U.S. Chamber of Commerce. *Congressional Digest* 7 (1928).

U.S. Department of Homeland Security, Federal Emergency Management Agency
(FEMA). "FEMA History" (www.fema.gov/about/history.shtml).

U.S. Environmental Protection Agency. "America's Wetlands: Our Vital Link
between Land and Water" (www.water.epa.gov/type/wetlands/types_index.cfm).

——. "Economic Benefits of Wetlands." EPA843-F-06-004, May 2006.

——. "Functions and Values of Wetlands," 2001 (www.epa.gov/owow/wetlands/
pdf/fun_val.pdf).

——. "Hydraulic Fracturing" (http://water.epa.gov/type/groundwater/uic/class2/
hydraulicfracturing/index.cfm).

——. "Watershed Academy: Wetland Functions and Values" (http://www.epa.gov/
watertrain/wetlands/index.htm).

——. "What are Wetlands?" (water.epa.gov/lawsregs/guidance/wetlands/
definitions.cfm).

——, Office of Environmental Justice. "Guidance to Assessing and Addressing
Allegations of Environmental Injustice: Working Draft," January 10, 2001.

U.S. Fish and Wildlife Service. "Holt Collier National Wildlife Refuge: Refuge History" (www.fws.gov/holtcollier/history.html).

U.S. Geological Survey. "About the Upper Mississippi River System," 2006
(www.umesc.usgs.gov/umesc_about/about_umrs.html).

——. "About Wetlands" (www.nwrc.usgs.gov/wetlands.htm).

——. "Kansas Floods" (http://ks.water.usgs.gov/pubs/fact-sheets/fs.041-01.html).

——. "Louisiana Water" (www.americaswetland.com/custompage.
cfm?pageid=4&cid=198).

——. "The Missouri River Story" (http://infolink.cr.usgs.gov/The_River/
MORstory.htm).

——. "Significant Floods of the 20th Century" (www.ks.water.usgs.gov/pubs/
fact-sheets/fs.024-00.table.gif).

Vargo, Cecile Page. "The Great Floods of the San Gabriel Mountains, Part III:
The Floods of 1938 and Beyond." *Explore Historic California*, March 2005
(www.explorehistoricalif.com/floods3.html).

"Warren County—2011, Comprehensive Development Plan (Land Use Plan)"
(www.warrencountync.com/_fileUploads/forms/144_WarrenCountyLandUse
Plan.pdf).

Water Science & Technology Board, Commission on Flood Control Alternatives
in the American River Basin. "Flood Risk Management and the American River
Basin: An Evaluation," 1995.

Weldon, Luci. "State Deeds PCB Landfill to County." *Virginia/North Carolina News*, April 21, 2010 (http://vancnews.com/articles/2010/04/22/warrenton/news/news52.txt).

Whayne, Jeannie. "Robert Edward Lee Wilson (1865–1933)." *Encyclopedia of Arkansas History and Culture*, 2009 (www.encyclopediaofarkansas.net/encyclopedia/entry-detail.aspx?entryID=1800).

White, Gilbert F. "Human Adjustment to Floods." *Research Paper* 29, Department of Geography, University of Chicago (1945).

White, Iain. "The Absorbent City: Urban Form and Flood Risk Management." *Proceedings of the Institution of Civil Engineers (ICE)—Urban Design and Planning* 4 (2008).

Wilkerson, Isabel. "Cruel Flood: It Tore at Graves, and at Hearts." *New York Times*, August 26, 1993.

World Resources Institute. "Banking on Nature's Assets: How Multilateral Development Banks Can Strengthen Development Using Ecosystem Services." Table 3, 2009 (www.wri.org).

Wright, James M. "The Nation's Responses to Flood Disasters: A Historical Account." *Association of State Floodplain Managers* (2000).

Yardley, Jonathan. "The Sage of Baltimore." *Atlantic Monthly*, December 2002, 129–40.

Zellmer, Sandra. "A New Corps of Discovery for Missouri River Management." *Nebraska Law Review* 83 (2004).

———. "Sustaining Geographies of Hope: Cultural Resources on Public Lands." *University of Colorado Law Review* 73 (2002).

———. "The Devil, the Details, and the Dawn of the 21st Century Administrative State: Beyond the New Deal." *Arizona State Law Journal* 32 (2000).

Zimmerman, Nikaela J. "The Flood—1903: Cafe Clock." *Kansas State Historical Society*, March 2008 (http://www.kshs.org/p/cool-things-cafe-clock/10286).

CASE LAW

Adolph v. Federal Emergency Management Agency, 854 F.2d 732 (5th Cir. 1988).

Alexander v. Sandoval, 532 U.S. 275, 287-88 (2001).

Alfred Oliver & Co. v. Board of Com'rs of Orleans Levee Dist., 169 La. 438, 125 So. 441 (1929).

Allison v. Barberry Homes, Inc., No. 982935, 2000 WL 1473121, at 3 (Mass. Super. Ct. 2000).

Arkansas v. Tennessee, 246 U.S. 158, 173 (1918).

Arkansas v. Tennessee, 247 U.S. 461 (1918).

Arkansas v. Tennessee, 269 U.S. 152 (1925).

Arkansas v. Tennessee, 397 U.S. 88 (1970).

Armstrong v. Francis Corp., 120 A.2d 4 (1956).

Big Oak Farms, Inc., v. United States, 105 Fed. Cl. 48 (2012).

Brown v. Board of Education, 347 U.S. 483 (1954).

Carland v. Aurin, 53 S.W. 940 (Tenn. 1899).

Central Green Co. v. United States, 531 U.S. 425 (2001) (abrogating *James*, 1986).

Cissna v. Tennessee, 242 U.S. 195 (1916).

Cissna v. Tennessee, 246 U.S. 289 (1918).

Cubbins v. The Mississippi River Commission, 241 U.S. 351 (1916).

Culbertson v. Knight, 152 Ind. 121, 52 N.E. 700 (1899).

Danforth v. United States, 308 U.S. 271 (1939).

Empire State Cattle Co. v. Atchison, T. & S. F. Ry. Co., 135 F. 135 (D. Kan. 1905), aff'd, 210 U.S. 1 (1908).

First English Evangelical Lutheran Church of Glendale v. County of Los Angeles, 482 U.S. 304 (1987).

First English Evangelical Lutheran Church of Glendale v. County of Los Angeles (Brief for Appellant), pp. 2–4, 1986WL727409 (1986).

First English Evangelical Lutheran Church of Glendale v. County of Los Angeles, 258 California Reporter 893 (Cal. App. 1989).

Foret v. Board of Levee Com'rs of Orleans Levee Dist., 169 La. 427, 125 So. 437 (1929).

Fryman v. United States, 901 F.2d 79 (7th Cir. 1990).

Gibbons v. Ogden, 22 U.S. (9 Wheat) 1 (1824).

Graci v. United States, 301 F. Supp. 947 (E.D. La., 1969), affirmed, 456 F.2d 20 (5th Cir. 1971), on remand, 435 F. Supp. 189 (E. D. La. 1977).

Gove v. Zoning Board of Appeals of Chatham, 831 N.E.2d 865 (Mass. 2005).

Hoffman Homes, Inc., v. EPA, 999 F.2d 256, 262 (7th Cir. 1993).

Hurley v. Kincaid, 285 U.S. 95 (1932).

In re Katrina Canal Breaches Consolidated Litigation, 471 F.Supp.2d 684 (E.D. La. 2007).

In re Katrina Canal Breaches Consolidated Litigation, 533 F.Supp.2d 615, 634 (E.D. La. 2008).

In re Katrina Canal Breaches Consolidated Litigation, 647 F. Supp. 2d 644 (E.D. La. 2009), aff'd, 673 F.3d 381 (5th Cir. 2012), reversed on rehearing, 2012 Westlaw 4343775, F.3d (5th Cir. Sept. 2012).

In re Operation of Missouri River System Litigation, 421 F.3d 618 (8th Cir. 2005), cert. denied, 126 S.Ct. 1879-1880 (2006).

Jackson v. U.S., 230 U.S. 1 (1913).

Kelo v. City of New London, 545 U.S. 469 (2005).

Leovy v. United States, 177 U.S. 621, 636 (1900).

Lingle v. Chevron U.S.A., Inc., 544 U.S. 528, 540 (transcript of oral argument, p. 21).

Louisiana Public Service Comm'n v. F.C.C., 476 U.S. 355 (1986).

Lucas v. South Carolina Coastal Council, 505 U.S. 1003 (1992).

Mammoth Oil Company, the Sinclair Crude Oil Purchasing Company and the Sinclair Pipe Line Company v. United States, 275 U.S. 13 (1927).

Marsh v. Oregon Natural Resources Council, 490 U.S. 360 (1989).

Matthews v. United States, 87 Ct. Cl. 662 (1938) and 113 F.2d 452 (8th Cir. 1940).

Minnesota Chippewa Tribe v. U.S., 230 Ct.Cl. 776 (1982).

Missouri v. Scott, 943 S.W.2d 730, 733 (Mo. App. 1997).

Moffatt Commission Co. v. Union Pac. R. Co., 113 Mo. App. 544, 88 S.W. 117 (1905).

NAACP v. Gorsuch, No. 82-768-CIV-5, slip op. at 10 (E.D.N.C. Aug. 1982).

National Mfg. Co. v. United States, 210 F.2d 263, 270-71 (8th Cir.), cert. denied, 347 U.S. 967 (1954).

North v. Johnson, 59 N.W. 1012 (Minn. 1894).

Palazzolo v. Rhode Island, 533 U.S. 606 (2001).

Palazzolo v. Rhode Island, No. WM-88-0297, 2005 WL 1645974 (R.I. Super. July 5, 2005), on remand from *Palazzolo v. Rhode Island*, 533 U.S. 606 (2001).

Palsgraf v. Long Island RR Co., 162 N.E. 99 (N.Y. 1928).

Pennsylvania Central Transportation Company v. City of New York, 438 U.S. 104 (1978).

Pennsylvania Coal v. Mahon, 260 U.S. 393 (1922).

Plessy v. Ferguson, 163 U.S. 537 (1896).

Public Lands Council v. Babbitt, 529 U.S. 728 (2000).

Ramold v. Clayton, 77 Neb. 178, 108 N.W. 980 (1906).

Roath v. Driscoll, 20 Conn. 533, 541 (1850).

Sabine River Authority v. U.S. Department of the Interior, 951 F.2d 669, 672 (5th Cir. 1992).

South Dakota v. Ubbelohde, 330 F.3d 1014, 1027 (8th Cir. 2003), cert. denied, 124 S.Ct. 2015 (2004).

Story v. Marsh, 732 F.2d 1375 (8th Cir. 1984).

Surocco v. Geary, 3 Cal. 69, 73 (1853).

Tucker v. Badoian, 384 N.E.2d 1195 (Mass. 1978).

Town of Sudbury v. Dep't of Pub. Utils., 218 N.E.2d 415, 424 (Mass. 1966).

United States v. Archer, 241 U.S. 119 (1916).

United States v. Caltex, Inc., 344 U.S. 149, 154 (1952), rehearing denied, 344 U.S. 919 (1953).

United States v. Carmack, 329 U.S. 230 (1946).

United States v. Carroll Towing Co., 159 F.2d 169 (2d Cir. 1947).

United States v. James, 760 F.2d 590 (5th Cir. 1985), rev'd, 478 U.S. 597 (1986).

United States v. James, 478 U.S. 597 (1986), citing S. Rep. No. 619, 70th Cong., 1st Sess. 12 (1928).

United States v. Ward, 676 F.2d 94 (4th Cir. 1982).

Wana the Bear v. Cmty. Constr., Inc., 180 Cal. Rptr. 423, 426 (Cal. Ct. App. 1982).

Watts v. Evansville, Mt. C. & N. Ry. Co., 120 N.E. 611 (Ind. App. 2. Div. 1918), superseded by *Watts v. Evansville, Mt. C. & N. Ry. Co.*, 129 N.E. 315 (Ind. 1921).

Yankton Sioux Tribe v. United States Army Corps of Engineers, 83 F. Supp. 2d 1047+ (D.S.D. 2000).

Yankton Sioux Tribe v. United States Army Corps of Engineers, 209 F. Supp. 2d 1008 (D.S.D. 2002).

STATUTES, LEGISLATIVE MATERIALS, REGULATIONS,
AND EXECUTIVE ORDERS

69 Cong.Rec. (1928): 5294 (statement of Sen. James Reed).

69 Cong.Rec. (1928): 7011 (statement of Rep. Robert Crosser).

15 U.S.C. § 2605(e); 40 C.F.R., Part 761.

25 U.S.C. § 3002.

25 U.S.C. § 3002(c); 43 C.F.R. § 10.4(d)(1).

25 U.S.C. § 3002(d); 43 C.F.R. § 10.4(b) (2001).

Act of June 28, 1879, 21 Stat. 37.

Act of Mar. 1, 1917, 64th Cong., Pub. Law No. 367.

Arkansas Statehood Act, 5 Stat. 50, June 15, 1836.

Civil Rights Act of 1964, 42 U.S.C. § 2000d.

Coastal Zone Management Act of 1972, 16 U.S.C. §§ 1451-56 (West 2006).

Code of Federal Regulations, volume 44, § 65.15 (NFIP regulations).

Control of Floods on Mississippi River, Senate Rep. No. 70-448 (1928).

Disaster Relief Act of 1950, Pub. L. No. 81-875, 64 Stat. 1109, 81 Cong. Ch. 1125, Sept. 30, 1950.

Exec. Order No. 12,898, 59 Feb. Reg. 32 (11 Feb. 1994).

Flood Control Act of 1928, 45 Stat. 534, 33 U.S.C. §§ 702a-m, May 15, 1928.

Flood Control Act of 1936, ch. 688, 49 Stat. 1570, codified at 33 U.S.C. § 701(a).

Flood Control Act of 1944, 58 Stat. 887, codified in various provisions of Titles 16, 33, and 43 of the U.S. Code.

Flood Control Act of 1965 (authorizing Lake Pontchartrain and Vicinity Hurricane Protection Project).

Flood Disaster Protection Act of 1973, Pub. L. No. 93-234, 87 Stat. 975 (Dec. 31, 1973).

Flood Insurance Reform Act, P.L. 108-264, 118 Stat. 712 (2004).

H.R. Rep. No. 1101, 70th Cong., 1st Sess. 13 (1928).

National Flood Insurance Act of 1968, 42 U.S.C. §§ 4001-4129.

National Flood Insurance Act of 1968, S. Rep. No. 93-583, p. 3219, Nov. 29, 1973.

National Flood Insurance Reform Act, P.L. 103-325, 108 Stat. 2255 (1993).

Public Law Number 84-1016, 70 Stat. 1078 (repealed 1957).

Public Law Number 103-325, 108 Stat. 2255, codified at 42 U.S.C. § 4013(c)(1).

Public Law Number 103-181, § 3 (1993), codified at 42 U.S.C. § 5170c Note.

Public Law No. 109-234 (de-authorizing Mississippi River-Gulf Outlet).

Ransdell-Humphreys Flood Control Act of Mar. 1, 1917, Pub. L. No. 64-367, 39 Stat. 948.

Restatement (Second) of Torts § 196 (1995).

Rivers and Harbors Act of 1907, ch. 2509, 34 Stat. 1110, Mar. 2, 1907.

Rivers and Harbors Act of 1930, Ch. 847, 46 Stat. 918, 927, 928, July 3, 1930.

Senate Report No. 93-583 (1973), reprinted in 1973 U.S.C.C.A.N. 3217, 3219.

Soil Conservation Act of 1935, Pub. L. No. 74-46, 49 Stat. 163 (codified as amended at 16 U.S.C. §§ 590a-590q-3).

South Carolina Beachfront Management Act, S.C. Code Ann. § 48-39-290(A).

Stafford Disaster Relief and Emergency Assistance Amendments of 1988, Pub. L. No. 100-707, 102 Stat. 4689 (1988) (codified at 42 U.S.C. §§ 5121-5202 (1988)). The 1988 amendments incorporated the *Robert T. Stafford Disaster Relief and Emergency Assistance Act of 1974,* Pub. L. No. 93-288, 88 Stat. 143 (1974).

Tennessee Statehood Act, 1 Stat. 491, June 1, 1796.

Water Resources Development Act of 2000, § 101(b)(18).

OTHER

1913 Flood Postcard Collection. Dayton, Ohio, Metro Library (http://content .daytonmetrolibrary.org/cdm4/browse.php?CISOROOT=%2Ffloodpost)

The Center for Land Use Interpretation. "Old River Control Structure," Photo Archive #LA3126 (http://ludb.clui.org/ex/i/LA3126/).

Coovert, J. C. "Floods," 2001 (www.jccoovert.com/floods/flood_gallery.html).

"Fatal Flood, Maps: Comparing Floods." Public Broadcast Service (www.bps.org/ wgbh/amex/flood/maps/index.html).

Fessler, Pam. "New Orleans Public Housing Slowly Evolving." *Weekend Edition Sunday,* National Public Radio, August 29, 2010 (www.npr.org/templates/story/ story.php?storyId=129448906).

"Flooded Missouri Farmers File Suit against Government." *All Things Considered,* National Public Radio, May 3, 2011.

Inskeep, Steve, and David Greene. "President Bush Returns to New Orleans." *Morning Edition,* National Public Radio. August 29, 2007 (www.npr.org/ templates/story/story.php?storyId=14016105&ps=rs).

"Losing Ground: The Disappearing Delta." *Bill Moyers' Journal,* September 6, 2002 (http://www.pbs.org/moyers/journal/archives/losingground_ts.html).

Siegal, Robert. "Letters: Cussing, Economy, Katrina." *All Things Considered,* National Public Radio, January 27, 2009 (www.npr.org/templates/story/story .php?storyId=99919399).

Sullivan, Bob. "Wetlands Erosion Raises Hurricane Risks." MSNBC.com, August 29, 2005 (http://www.nbcnews.com/id/9118570/ns/technology_and_science -science/t/wetlands-erosion-raises-hurricane-risks/).

Survey USA. "Approval Ratings for All 50 Governors; Profile: Ray Nagin." *BBC News,* May 21, 2006 (www. news.bbc.co.uk/2/hi/americas/4623922.stm).

"Why Societies Collapse: Jared Diamond at Princeton University." Australian Broadcasting Corporation, October 27, 2002 (http://www.abc.net.au/radionational/programs/backgroundbriefing/why-societies-collapse-jared-diamond-at-princeton/3526390).

INDEX

ABOUT THE AUTHORS

CHRISTINE A. KLEIN is the Chesterfield Smith Professor of Law at the University of Florida Levin College of Law, where she directs the law school's advanced degree program in environmental and land-use law. Klein specializes in water law, natural resources law, and property law. She recently served as a member of the National Academy of Sciences Committee on Sustainable Water and Environmental Management in the California Bay-Delta. Previously, she was a western water lawyer in the Office of the Colorado Attorney General, and also clerked for the Honorable Richard P. Matsch at the U.S. District Court in Denver.

SANDRA B. ZELLMER holds the Robert B. Daugherty Chair at the University of Nebraska College of Law. Zellmer specializes in water and natural resources law, environmental law, and torts. She is active in the American Bar Association, and she recently served as a member of the National Academy of Sciences Committee on Missouri River Recovery. Previously, Zellmer was an attorney for the U.S. Department of Justice Environment and Natural Resources Division. She also practiced law at Faegre & Benson in Minnesota and clerked for the Honorable William W. Justice at the U.S. District Court in Tyler, Texas.